Collins

Student Support Materials for

AQA

AS BIOLOGY

SPECIFICATION (A)

Module 3: **Pathogens and Disease**

Mike Boyle
Series consultant: Bill Indge

This booklet has been designed to support the AQA (A) Biology AS specification. It contains some material which has been added in order to clarify the specification. The examination will be limited to material set out in the specification document.

Published by HarperCollins*Publishers* Limited
77–85 Fulham Palace Road
Hammersmith
London
W6 8JB

www.**Collins**Education.com
Online support for schools and colleges

First published 2001

ISBN 0 00 327708 9

Mike Boyle asserts the moral right to be identified as the author of this work

British Library Cataloguing in Publication Data
A catalogue record for this publication is available from the British Library

Cover designed by Chi Leung
Editorial, design and production by Gecko Limited, Cambridge
Printed and bound by Scotprint, Haddington

The publisher wishes to thank the Assessment and Qualifications Alliance for permission to reproduce the examination questions.

Other useful texts

Full colour textbooks
Collins Advanced Modular Sciences: Biology AS
Collins Advanced Science: Biology

Student Support Booklets
AQA (A) Biology: 1 Molecules, Cells and Systems
AQA (A) Biology: 2 Making Use of Biology

To the student

What books do I need to study this course?

You will probably use a range of resources during your course. Some will be produced by the centre where you are studying, some by a commercial publisher and others may be borrowed from libraries or study centres. Different resources have different uses – but remember, owning a book is not enough – it must be *used.*

What does this booklet cover?

This *Student Support Booklet* covers the content you need to know and understand to pass the module test for AQA (A) Biology AS Module 3: Pathogens and disease. It is very concise and you will need to study it carefully to make sure you can remember all of the material.

How can I remember all this material?

Reading the booklet is an essential first step – but reading by itself is not a good way to get stuff into your memory. If you have bought the booklet and can write on it, you could try the following techniques to help you to memorise the material:

- underline or highlight the most important words in every paragraph
- underline or highlight scientific jargon – write a note of the meaning in the margin if you are unsure
- remember the number of items in a list – then you can tell if you have forgotten one when you try to remember it later
- tick sections when you are sure you know them – and then concentrate on the sections you do not yet know.

How can I check my progress?

The module test at the end is a useful check on your progress – you may want to wait until you have nearly completed the module and use it as a mock exam or try questions one by one as you progress. The answers show you how much you need to do to get the marks.

What if I get stuck?

A colour textbook such as *Collins Advanced Modular Sciences: Biology AS* provides more explanation than this booklet. It may help you to make progress if you get stuck.

Any other good advice?

- You will not learn well if you are tired or stressed. Set aside time for work (and play!) and try to stick to it.
- Don't leave everything until the last minute – whatever your friends may tell you, it doesn't work.
- You are most effective if you work hard for shorter periods of time and then take a (short!) break. 30 minutes of work followed by a five or ten minute break is a useful pattern. Then get back to work.
- Some people work better in the morning, some in the evening. Find out which works better for you and do that whenever possible.
- Do not suffer in silence – ask friends and your teacher for help.
- Stay calm, enjoy it and ... good luck!

Further explanation references give a little extra detail, or direct you to other texts if you need more help or need to read around a topic.

There are rigorous definitions of the main terms used in your examination – memorise these exactly.

AQA (A) Biology AS

12.2 The parasites responsible for malaria and schistosomiasis show structural and physiological adaptations that enable them to infect new hosts and survive inside them

Parasites and parasitism

A parasite is an organism that lives in or on a host organism. It gains a nutritional advantage from this relationship while the host suffers a disadvantage.

From the definition it would seem that bacteria and viruses are parasites, but the term is normally reserved for single-celled and multicellular *animals.*

Parasites show the following features:

- they can survive and reproduce inside the host
- they show parasitic degeneration. Compared to their free-living relatives, parasite species often show a reduction in complexity; they lack organs of locomotion, sense organs or even a gut. If a parasite is surrounded by digested food, there is no need to go in search of it – it simply absorbs the digested food molecules through its body wall
- their life cycle is adapted so they can multiply quickly inside the host and spread to new hosts. The reproductive capacity of parasites is vast, because many individuals will not complete their life cycle as it depends to a large extent on luck.

The word malaria comes from the Italian for 'bad air', and indicates that malaria is mostly found in areas with many stagnant pools and ditches where mosquitoes breed.

Malaria and schistosomiasis are probably the two most serious parasitic diseases in the world today, accounting for millions of deaths per year in tropical areas. Recent estimates suggest that there are over 500 million cases of malaria world-wide, 90% of them in Africa. Up to 2.7 million people are killed annually, 1 million of them African children under 5 years old.

Malaria

Malaria is caused by infection with the microorganism *Plasmodium*. There are four different species, all with the same basic life cycle and mode of transmission, via the female *Anopheles* mosquito. One of the key survival strengths of this parasite is its ability to avoid the host's immune system, which it achieves by invading the host's cells – see the life cycle in Fig 6.

The key features of the *Plasmodium* life cycle include:

- the mosquito becomes infected with the parasite when it bites an infected person
- *Plasmodium* produces gametes that fuse inside the mosquito
- thousands of immature malarial parasites are produced by mitosis
- the infective stage of the parasite invades the mosquito's salivary glands
- the mosquito bites an uninfected person, injecting the parasite
- the infective stage travels in the blood to the liver, where the parasites multiply again

Do not state that the mosquito causes malaria – the mosquito is the vector, or carrier.

10

The examiner's notes are always useful – make sure you read them because they will help with your module test.

The main text gives a very concise explanation of the ideas in your course. You must study all of it – none is spare or not needed.

Table 1
Outline of this module

Section	Basic contents	Revision complete
12.1	Bacteria and viruses as examples of pathogens	
12.2	Two parasites responsible for malaria and schistosomiasis (bilharzia)	
12.3	The mammalian immune system	
12.4	Cell division: mitosis and meiosis	
12.5	Genes, nucleic acids and protein synthesis	
12.6	Gene technology can be used to combat disease	
12.7	Heart disease and cancer	
12.8	Diagnostic techniques	
12.9	Drugs used in the control/treatment of disease	

12.1 Bacteria and viruses are examples of pathogenic microorganisms

The differences between prokaryotic and eukaryotic cells are covered in Module 1, but may be examined in this module test, so make sure you have learnt them.

Bacteria

Bacteria are simple, single-celled organisms (Fig 1). Their cells are described as **prokaryotic**; they are small, with very few structures (organelles) inside. In contrast, virtually all other organisms – plants, animals, fungi, etc. – are **eukaryotic**; they have large, relatively complex cells.

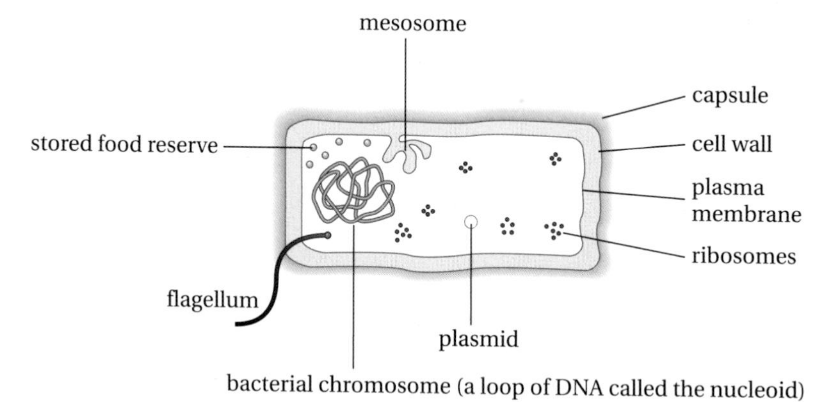

Fig 1
The basic structure of a bacterium

Bacteria can grow at a phenomenal rate in conditions that are:

- moist
- at a suitable temperature (usually in the region of 20–45 °C). Growth is inhibited at low temperature, and bacteria are killed by high temperature
- with plenty of nutrients – bacteria need some carbohydrate or other source of carbon, a supply of nitrogen and a few mineral salts. Bacteria also obtain nutrients by secreting enzymes that can digest some of the surrounding substances.

Given the right conditions – such as in culture or inside the human body – a bacterial population will grow in a clearly defined pattern (Fig 2). This is known as a **sigmoid growth curve** (after the Greek letter sigma) and consists of the following phases:

- **lag phase** – growth is initially slow; the bacteria might not be able to digest and absorb the growing medium. They may need to make and secrete new enzymes and this takes time because it requires the activation of genes and the synthesis of proteins
- **log or exponential phase** – growth is rapid because there are no **limiting factors**. Growth is by **binary fission**; once the cell reaches a certain critical size, it divides in two. In optimum conditions, the population can double in as little as 20 minutes
- **stationary phase** – growth levels out as one or more limiting factors begin to have an effect. A limited supply of food or an accumulation of waste are two common factors that limit bacterial growth
- **decline phase** – conditions become increasingly limiting; nutrients run out and waste builds up. Bacteria cannot survive in such circumstances.

Exponential growth means that the population doubles with each generation. Some organisms show super-exponential growth, when the population more than doubles each generation, e.g. when mice have several litters of six or more babies.

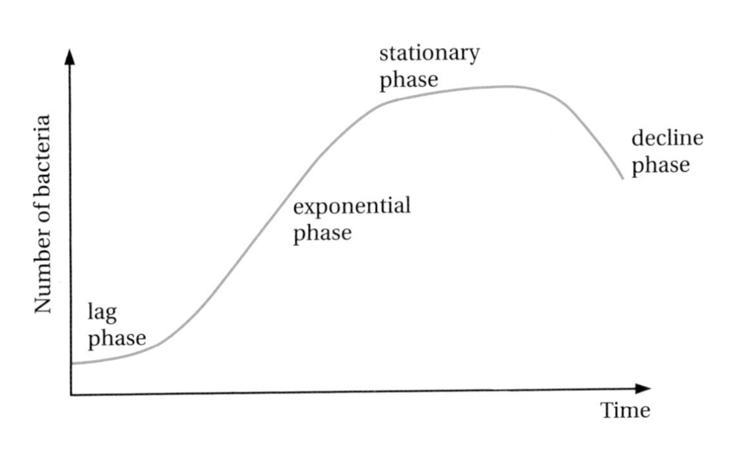

Fig 2
Standard bacterial growth curve

The population growth of a microorganism can be measured by estimating the number of individuals *per unit volume*, e.g. per mm^3 per hour. One way to do this is to use a modified microscope slide called a **haemocytometer** (Fig 3).

Viruses

Viruses are a major cause of disease, but there is still debate as to whether they should be classed as 'living' at all. They do not feed or respire, and cannot reproduce without invading a host cell.

All viruses consist of nucleic acid (DNA or RNA) inside a protein coat, but they show considerable variation around this basic structure. Once inside a cell, the virus takes over the host's organelles and biochemical processes, notably protein synthesis, causing the cell to make more virus particles.

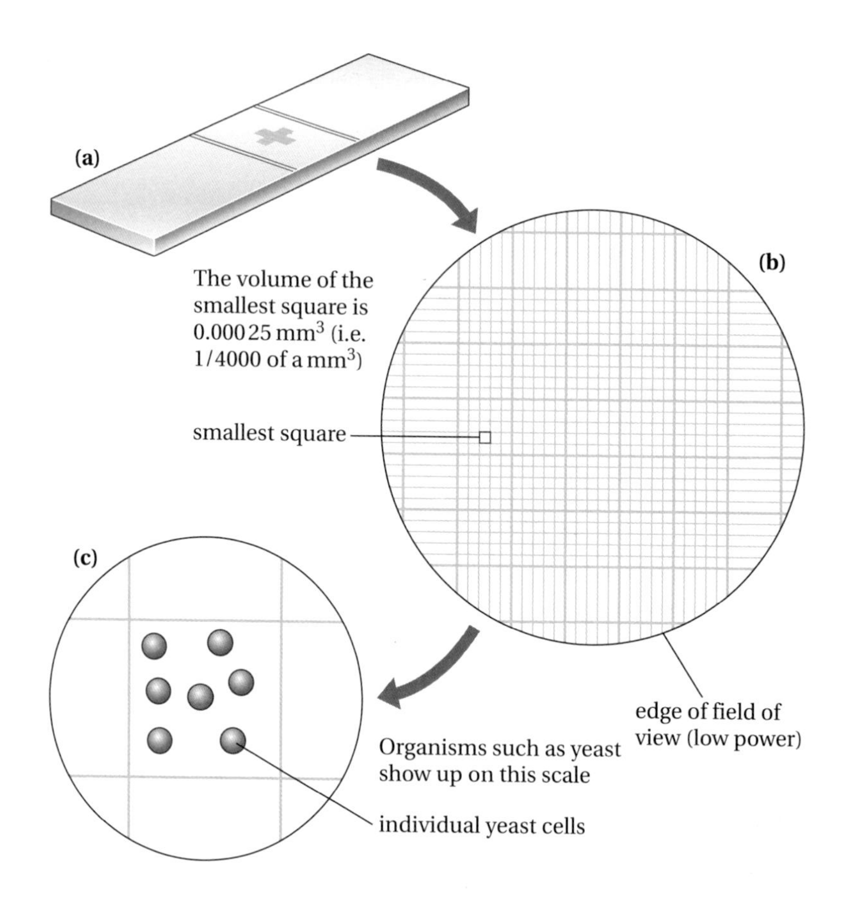

Fig 3
A haemocytometer is a device for counting red blood cells, but it can also be used for counting bacteria or yeast cells. The volume of the fluid below each square is known, so the population of microorganisms can be worked out. For example, if there are, on average, 12 bacteria per square, and the volume of each square is 0.000 25 mm^3, the population can be said to be 12/0.000 25 = 48 000 bacteria per mm^3. **(a)** Whole slide showing position of grid **(b)** central part of grid **(c)** smallest square

Within a short time millions of viruses can be made inside a single cell. They burst out of the cell, destroying it in the process and releasing various toxins into the bloodstream. Each new virus has the potential to invade other host cells and repeat this **lytic cycle**.

This specification requires you to know about the structure (Fig 4) and replication (Fig 5) of the human immunodeficiency virus (HIV). This virus causes AIDS (acquired immunodeficiency syndrome) because it destroys certain essential cells in the immune system, reducing its effectiveness. Eventually, most infected individuals lose the ability to control secondary infections that would normally cause few problems.

HIV is a comparatively complex virus. Fig 4 shows the essential features that include:

- a core that contains RNA; for this reason HIV is classed as a **retrovirus**
- the enzyme reverse transcriptase, which makes DNA from RNA, i.e. transcription in reverse, hence the name retrovirus (retro = reverse/backwards)
- an outer lipid membrane that allows it to enter cells easily by fusing with the plasma membrane
- glycoprotein spikes that can change from one generation to the next, making it very hard for the immune system to recognise the virus.

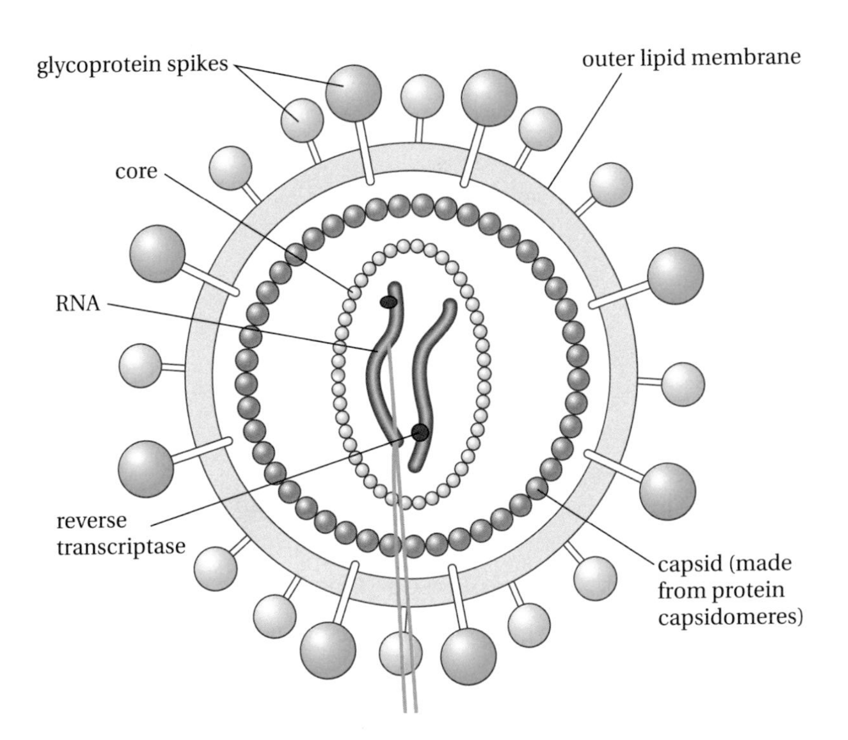

Fig 4
Structure of the human immunodeficiency virus (HIV)

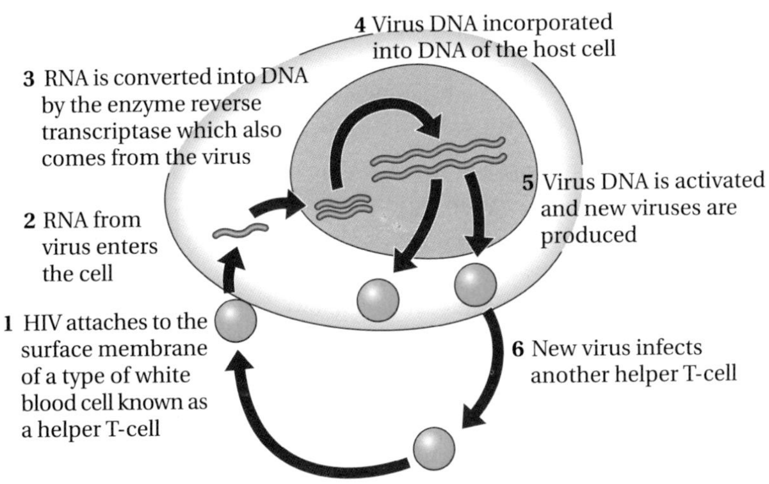

Fig 5
The life cycle of HIV

The association of microorganisms with disease

The invention and refinement of the microscope was a vital breakthrough in our understanding of the causes of disease. For the first time, we could see bacteria and investigate the causes of infectious disease scientifically.

One of the pioneers of bacteriology was Robert Koch. **Koch's postulates** are basic rules which state that a disease can be said to be caused by a particular microorganism if:

1 the microorganism can always be found in diseased individuals

2 the microorganism taken from the host can be grown in pure culture

3 samples from the culture produce the disease when injected into a new, healthy host

4 the newly infected host yields new, pure cultures of the microorganism identical to those obtained in step 2.

How bacteria cause disease

Once a bacterium has entered the body, it must do several things before it can cause disease symptoms. It must:

- attach to the host's cells
- penetrate the host's cells
- multiply in the tissues of the host
- evade the host's immune system.

The symptoms of the disease are caused in two main ways: the host's cells are damaged by the bacteria, and the by-products of bacterial metabolism, **exotoxins**, interfere with the host's metabolism.

The ability of bacteria to cause disease depends on:

- **situation** – when bacteria get into parts of the body that are normally sterile, e.g. through cuts
- **infectivity** – how many individual bacteria are needed to cause an infection. Some bacteria, e.g. *Salmonella enteritidis*, which causes food poisoning, have a low infectivity and many individuals have to be present before they cause symptoms. In contrast, *Salmonella typhi*, which causes typhoid, is highly infectious and only a few individuals are needed before symptoms start
- **invasiveness** – how easily the bacteria can penetrate the host's cells, break into the circulation and spread around the body
- **pathogenicity** – the toxicity of substances excreted by the bacteria. Some are mild and produce few symptoms, others – such as the diphtheria toxin – are highly pathogenic and life-threatening.

Salmonella

spp. after a name, as in *Salmonella* spp., means two or more different species all belonging to the same genus. Scientific names always start with the name of the genus (with a capital letter) followed by the species name (no capital letter), e.g. *Salmonella typhi*.

***Salmonella* spp.** are food-poisoning bacteria. The genus *Salmonella* is very widespread, being found in a high proportion of eggs and uncooked chickens, for example. However, they do not cause symptoms – vomiting, stomach cramps, diarrhoea, etc. – unless many individuals are eaten. Basic food hygiene precautions reduce the incidence of salmonella food poisoning.

In contrast, *Salmonella typhi* is a highly dangerous bacterium that causes typhoid. Infection is usually by water or food contaminated with human faeces. Symptoms include a high fever, a rash and inflammation of the bowel. *Salmonella typhi* is highly invasive, and only a few organisms are needed to set up infection.

Tuberculosis (TB)

TB is a bacterial disease of the lungs. It is a major problem throughout the world, and is making a comeback in developed countries. TB is caused by two species: ***Mycobacterium tuberculosis*** and ***Mycobacterium bovis***.

TB facts:

- it is the biggest killer disease in the world – around 3 million deaths in 1998
- it is the commonest cause of death among AIDS patients, where TB is an *opportunistic* infection, taking advantage of the individual's weakened immune system
- it is transmitted via airborne droplets, or (more rarely) via unpasteurised milk
- TB is common in poor/socially disadvantaged areas, where people who sleep close together in large numbers are particularly at risk. TB is very common among the homeless
- it primarily affects the lungs, with secondary infections in the lymph nodes, bones and gut
- symptoms include a persistent cough, blood-smeared sputum, shortness of breath, fever and, in the long term, weight loss.

Once a major source of death in the UK, TB has been on the decline since World War II, mainly because of better housing and the availability of antibiotics. In recent years, however, TB has made a comeback as a result of various factors, including a rise in the number of homeless people, the global spread of AIDS and the emergence of antibiotic-resistant strains of *Mycobacterium tuberculosis.*

Once TB infection is confirmed (by analysis of sputum samples), treatment usually consists of isolation of the patient during the infective stage (2–4 weeks) and the use of a combination of drugs.

Drug treatment of TB takes several months. It is vital that people take the full course of prescribed drugs and do not stop just because they feel better, because mycobacteria are slow-growing and it may take months to kill them all. If treatment stops too soon, the more drug-resistant strains survive to continue the infection. This speeds up the development of resistant strains – a major problem throughout the world.

12.2 The parasites responsible for malaria and schistosomiasis show structural and physiological adaptations that enable them to infect new hosts and survive inside them

Parasites and parasitism

D

***A parasite** is an organism that lives in or on a host organism. It gains a nutritional advantage from this relationship while the host suffers a disadvantage.*

From the definition it would seem that bacteria and viruses are parasites, but the term is normally reserved for single-celled and multicellular *animals.*

Parasites show the following features:

- they can survive and reproduce inside the host
- they show parasitic degeneration. Compared to their free-living relatives, parasite species often show a reduction in complexity; they lack organs of locomotion, sense organs or even a gut. If a parasite is surrounded by digested food, there is no need to go in search of it – it simply absorbs the digested food molecules through its body wall
- their life cycle is adapted so they can multiply quickly inside the host and spread to new hosts. The reproductive capacity of parasites is vast, because many individuals will not complete their life cycle as it depends to a large extent on luck.

The word malaria comes from the Italian for 'bad air', and indicates that malaria is mostly found in areas with many stagnant pools and ditches where mosquitoes breed.

Malaria and schistosomiasis are probably the two most serious parasitic diseases in the world today, accounting for millions of deaths per year in tropical areas. Recent estimates suggest that there are over 500 million cases of malaria world-wide, 90% of them in Africa. Up to 2.7 million people are killed annually, 1 million of them African children under 5 years old.

Malaria

Malaria is caused by infection with the microorganism *Plasmodium.* There are four different species, all with the same basic life cycle and mode of transmission, via the female *Anopheles* mosquito. One of the key survival strengths of this parasite is its ability to avoid the host's immune system, which it achieves by invading the host's cells – see the life cycle in Fig 6.

The key features of the *Plasmodium* life cycle include:

- the mosquito becomes infected with the parasite when it bites an infected person
- *Plasmodium* produces gametes that fuse inside the mosquito
- thousands of immature malarial parasites are produced by mitosis
- the infective stage of the parasite invades the mosquito's salivary glands
- the mosquito bites an uninfected person, injecting the parasite
- the infective stage travels in the blood to the liver, where the parasites multiply again

E

Do not state that the mosquito *causes* malaria – the mosquito is the vector, or carrier.

- the parasites leave the liver and enter red blood cells
- the cycle is completed when another mosquito bites the host.

Fig 6
The life cycle of the malarial parasite

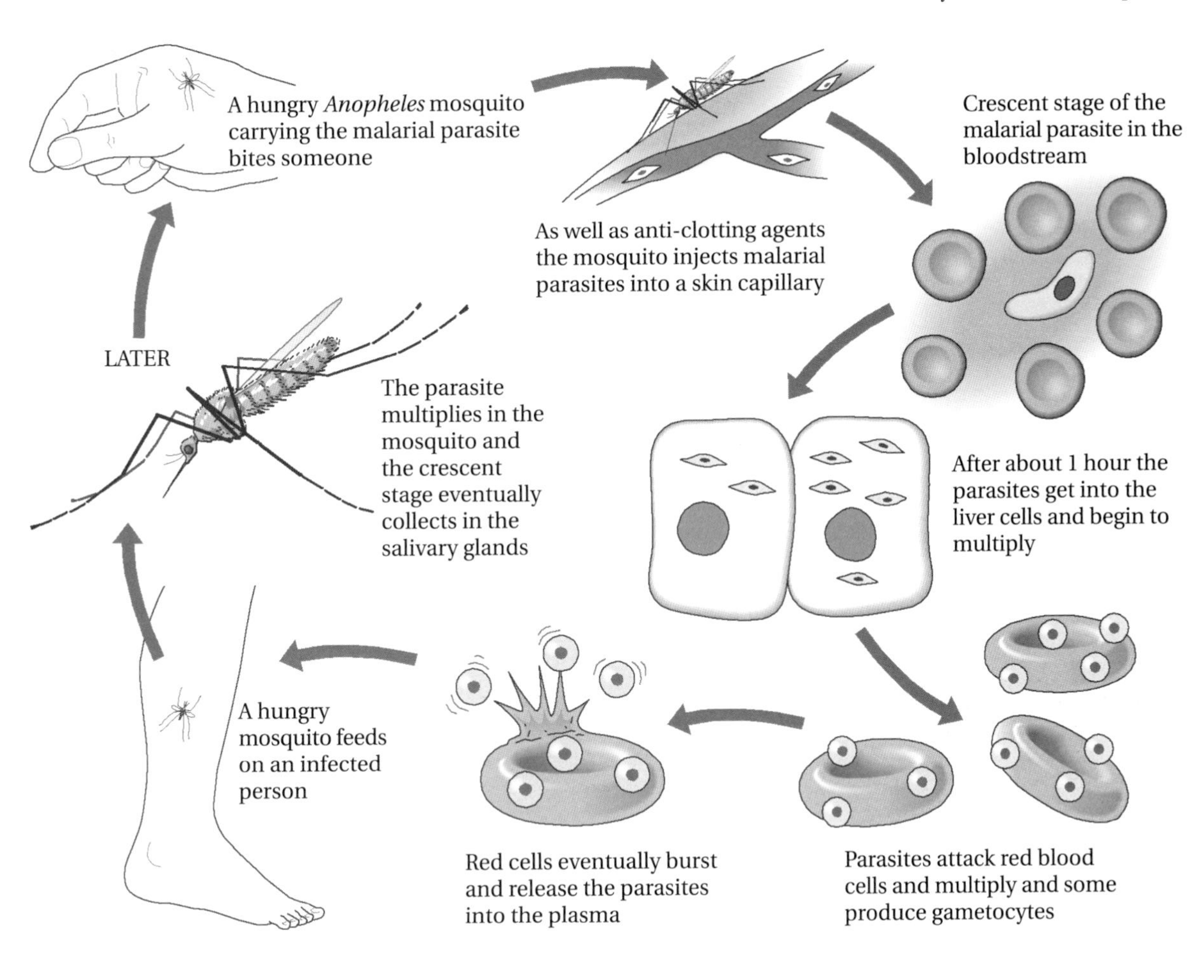

Schistosomiasis

This disease – also known as **bilharzia** – is caused by a schistosome parasite, or blood fluke. Flukes are multicellular animals related to tapeworms and flatworms.

The life cycle of the schistosome is shown in Fig 7.

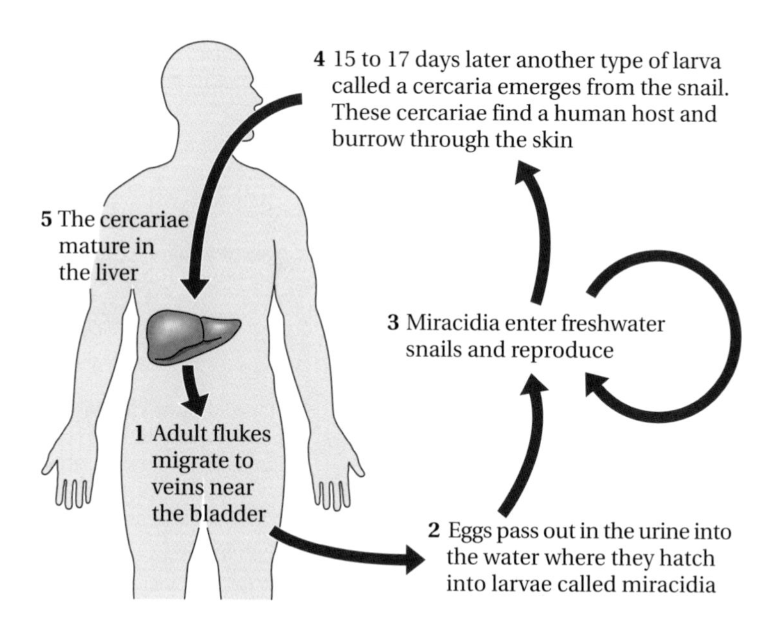

Fig 7
Schistosomiasis and the cycle of infection. Part of the life cycle takes place in humans and part in water snails. The snails are very difficult to kill and bilharzia is a problem wherever the snails are found

E The adaptations of parasites – as given in Table 2 – is a common topic for exam questions.

Bilharzia facts:

- the parasite infects humans through contact with water that is infested with water snails. The infective larvae of the parasite – called cercariae – come out of the snail and can burrow straight through human skin; once inside the human, they migrate to the liver where they mature
- adult flukes travel to the veins of the bladder where they attach and lay eggs; small infestations produce no symptoms, but large populations cause blockages and heavy bleeding
- the water snail is found in almost all waterways. Bilharzia can be avoided by staying away from infected water, but often this is impossible as some places only have a single source of water for washing, bathing and drinking
- the irrigation of farmland increases the range of the snail and of the disease.

Table 2
Adaptations of the malaria and bilharzia parasite compared

Adaptation	*Plasmodium*	*Schistosoma*
Attaching to the host	The malarial parasite spends much of its adult life inside the cells of its host. Attachment is not a problem.	Adult flukes live in the veins around the bladder. Because they are small they could easily be washed away by the blood. Suckers help them attach to the walls of the veins.
Resisting host defence mechanisms	The blood is a dangerous place for any parasite because of the white blood cells of the immune system (see Section 12.3). Within an hour of entering the body, malaria parasites are inside liver cells. Here they hide from the immune system. The antigens on the surface of a malarial parasite change frequently and the immune system cannot keep up.	Adult flukes have a remarkable way of avoiding the host's immune system: they disguise themselves by covering themselves with molecules from the host's red blood cells.
Reproduction and reproductive organs	There are many reproductive stages in the life cycle of the malarial parasite. The parasites reproduce in the mosquito, but this insect is small compared to a human. Once in their main host, malaria parasites also reproduce in the liver and in the blood cells, causing a rapid increase in parasitic load.	The life cycle of this fluke depends very heavily on chance. A person must urinate into fresh water; the miracidia larvae that hatch from the eggs must find a snail host; and the cercariae must encounter a human shortly after emerging from the snail. The production of large numbers of offspring is therefore the best way of ensuring survival. Each egg laid by an adult fluke can produce up to 200 000 cercariae.
The life cycle	Usually, blood from one person does not mix with that of someone else. So malaria parasites cannot depend on person-to-person contact for transmission. A second host is used. In malaria, the female *Anopheles* mosquito is the organism. When she bites someone with malaria, she takes in infected red blood cells. These parasites complete their life cycle in her body, finally migrating to her salivary glands. When she bites someone else, she injects a small amount of infected saliva, and malaria is passed on.	Part of the life cycle of *Schistosoma* is spent inside its human host; part is spent inside a freshwater snail. In many parts of the tropics where schistosomiasis is endemic, humans spend much of their time around areas of fresh water. They use it for washing clothes, irrigating crops, watering animals, removing waste and bathing. This increases the chances of the parasite being passed from one person to another.
Reduction of body systems	The malaria parasite spends most of its life inside one of its host's red blood cells. It does not have to move to find food, and the medium in which it lives has the same water potential as its own cytoplasm. It has therefore lost the capacity to locomote and to regulate cell water content – features of many free-living single-celled organisms.	Schistosomes do not have a complex nervous system because the conditions they find in the blood are remarkably constant. They cannot move either: a locomotory system is also unnecessary since the parasites' survival depends on it being anchored in the host.

12.3 Mammalian blood possesses a number of defensive functions

General mechanism of defence against disease

The body defends itself against invading microorganisms by putting up barriers to keep them out, and then having mechanisms to deal with those that do enter.

An overview of the body's defences is shown in Fig 8.

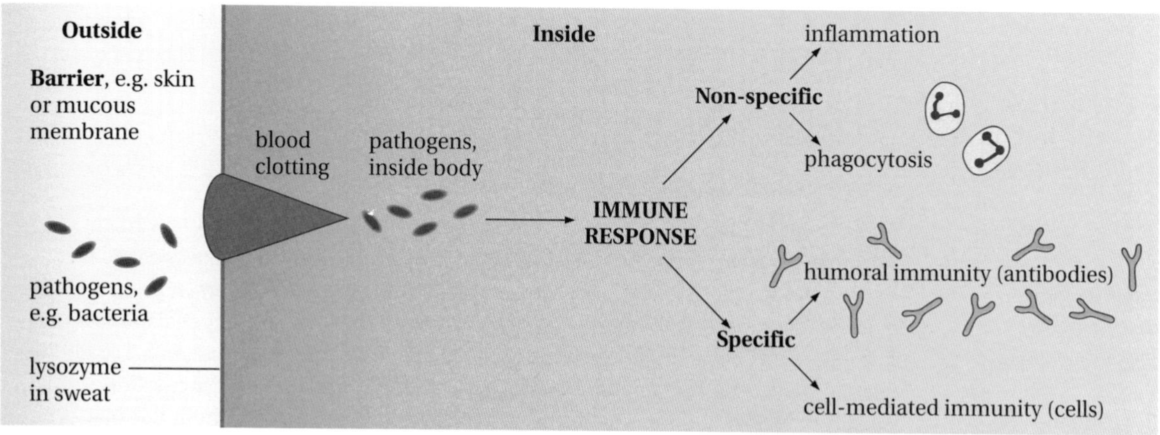

Fig 8
An overview of the body's defences: non-specific responses are general responses to damage. They include inflammation and phagocytosis of debris. Specific responses are targeted against individual types of microorganism

Blood clotting

Blood clotting enables the body to prevent blood loss. It also provides a physical covering to a wound that prevents the entry of microorganisms. It is important that blood only clots when it should because when a blood clot, a **thrombus**, blocks a vital vessel (e.g. in the heart or brain) the results can be fatal.

Blood clotting is a complex cascade reaction in which the activation of one molecule will lead to the activation of many more.

The following are essential components of blood clotting (Fig 9):

- **thromboplastin** is released by platelets and damaged blood vessel walls; it converts prothrombin into thrombin
- **prothrombin** is an inactive plasma protein; when converted into the active form, thrombin, it activates fibrinogen
- **plasma enzymes** are a variety of proteins, or factors, that contribute to the cascade reaction of blood clotting. If any one of these proteins is missing the clotting mechanism will not operate efficiently, e.g. haemophilia is caused by the lack of **factor VIII**
- **calcium ions** are essential for the conversion of fibrinogen to fibrin
- **fibrinogen** is a soluble plasma protein made by the liver. In the presence of calcium ions, thrombin and the essential factors, fibrinogen becomes **fibrin**, which is sticky and insoluble, so it forms a mesh. Red blood cells become entangled in the mesh and a clot forms.

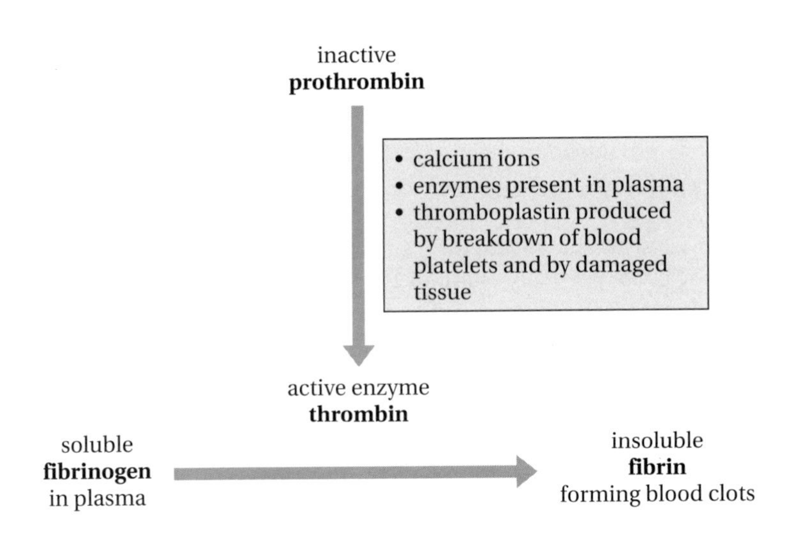

Fig 9
The blood clotting process

Phagocytosis

In this process, the most common type of white cells, the **neutrophils**, engulf any foreign material that has entered the body, e.g. small particles of dust in the lungs, or bacteria at the site of infection. Fig 10 shows the essential stages of phagocytosis.

Fig 10
Phagocytosis

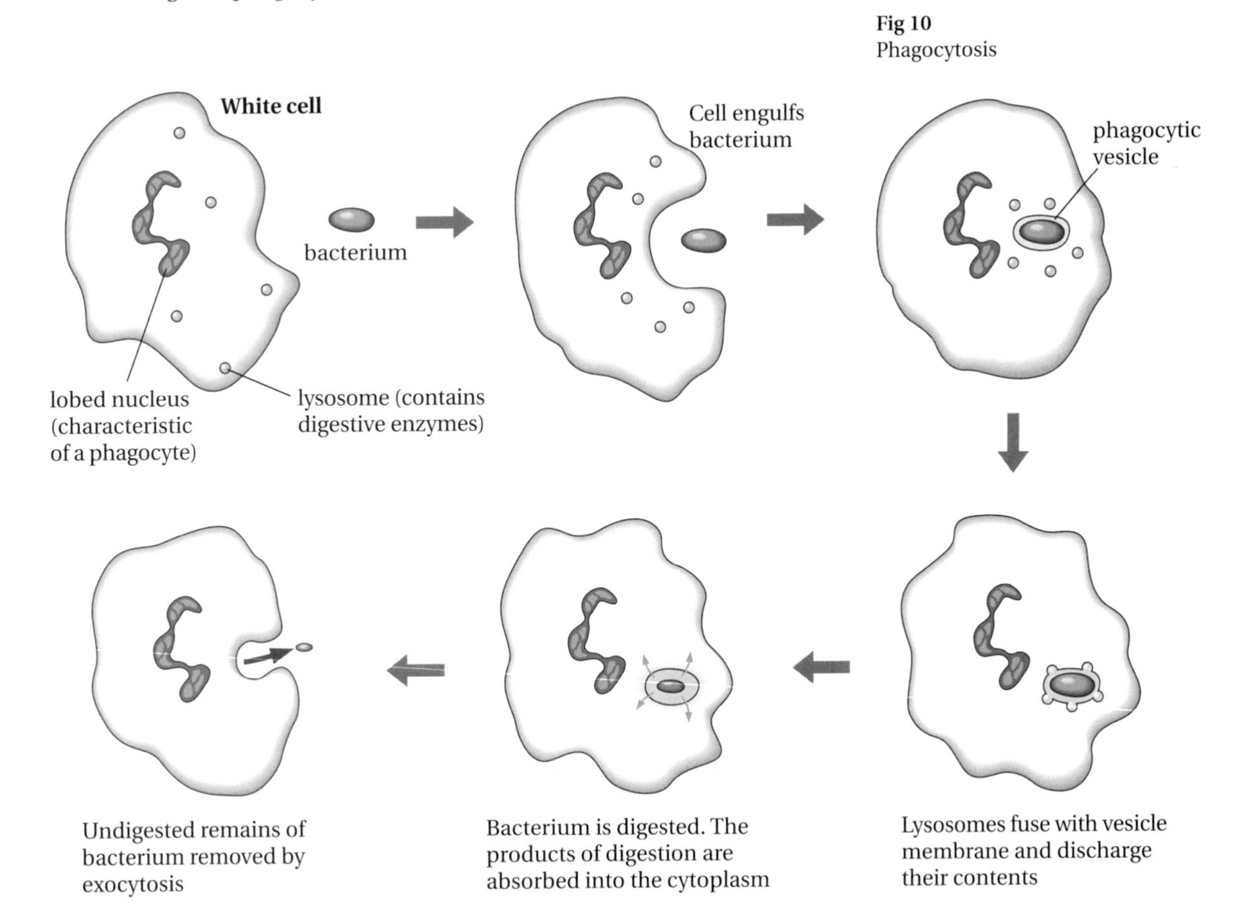

Principles of immunology

This aspect of immunity is often called *specific immunity* because it involves the recognition of specific pathogens. For example, if the polio virus gets into the body, the immune system will respond by producing specific anti-polio antibodies.

E Many candidates lose marks because they cannot distinguish between antigen and antibody.

In examination answers about the immune system there is too much talk of 'fighting'.

D ***An* antigen *is a substance – usually a protein or carbohydrate – that is not normally found in the host's body. The outer surface of any pathogen is recognised as foreign because it is made up of many antigens. Antigens stimulate the production of corresponding antibodies.***

An* antibody *is a protein made by the host's B-cells in response to a particular antigen. The antibody can combine with the antigen, and in some way neutralises the pathogen.

There are two types of specific immune response: **cell mediated** and **humoral** (Fig 11). Both involve the recognition of specific antigens and the production of tailor-made antibodies:

- in cell-mediated immunity the *whole cell* attacks the pathogen; cell mediated means 'carried out by the cells'
- in humoral immunity the *antibodies produced by the cell* attack the pathogen. Humoral means 'of the fluid' and refers to the fact that the antibodies in the plasma (fluid) attack the pathogen.

Fig 11
The difference between humoral and cell-mediated immunity

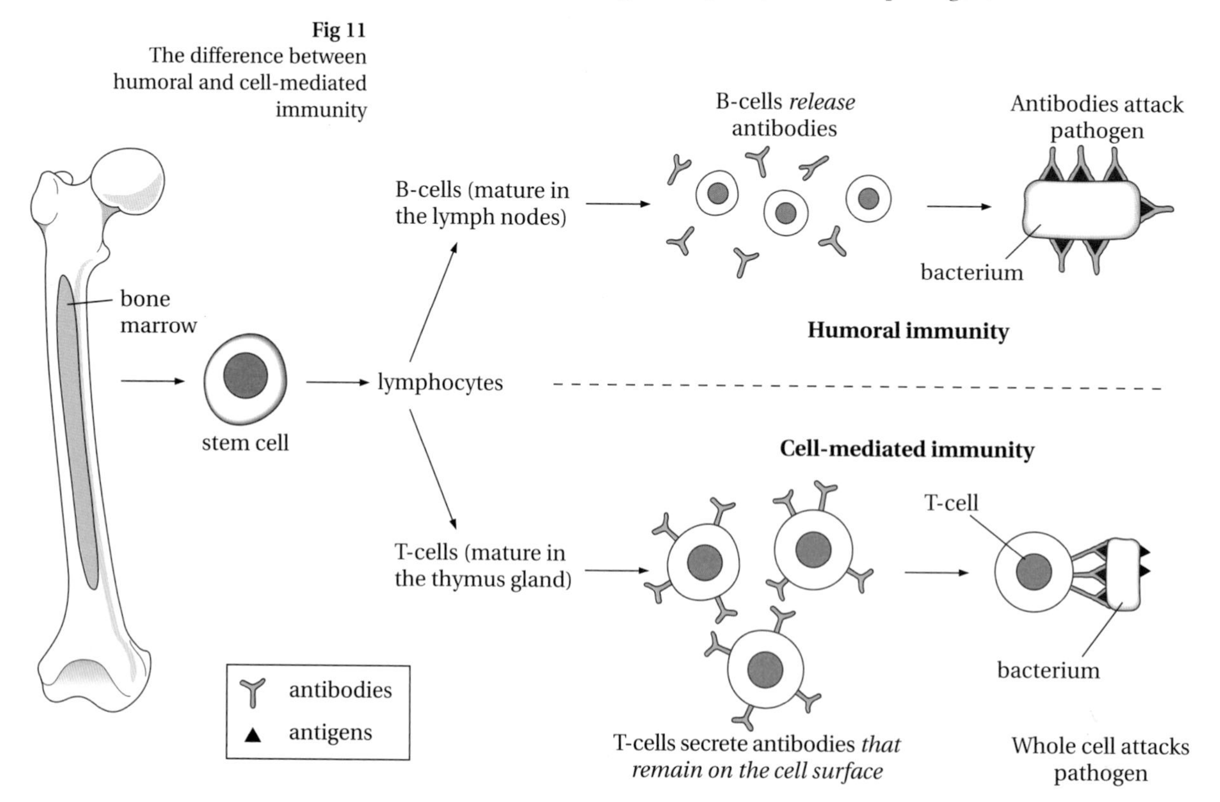

The primary and secondary immune response

The immune response involves the production of particular antibodies in response to a pathogen/antigen. There are two phases: primary and secondary.

In the **primary response**, **B lymphocytes** (or **B-cells**) – one type of white blood cell – are made and mature in the bone marrow (hence the B). It is thought that at birth there are thousands of small populations – or clones – of B cells, each one being capable of producing a particular antibody. When a pathogen (or antigen) gets into the body for the first time, it stimulates the relevant clone to multiply rapidly into:

- **plasma cells** – these short-lived cells circulate in the blood and secrete large amounts of antibodies. Plasma cells are all clones of the same cell, so all make the same antibody
- **memory cells** – these long-lived cells may exist for many years.

White blood cells are sometimes just called white cells because many of them spend most of their time out of the blood. White cells are also generally known as leucocytes, while red blood cells are known as erythrocytes.

The primary immune response may not be strong enough to prevent disease symptoms completely. Once the memory cells are in place, however, any further exposure to the antigen will stimulate the **secondary response**, in which the memory cells rapidly multiply into a large population of plasma cells, which can produce enough antibody to prevent infection before symptoms develop. The difference between the primary and the secondary response is seen in Fig 12. The second exposure to an antigen results in a faster, greater and more prolonged production of antibodies.

Fig 12
The primary and secondary immune response

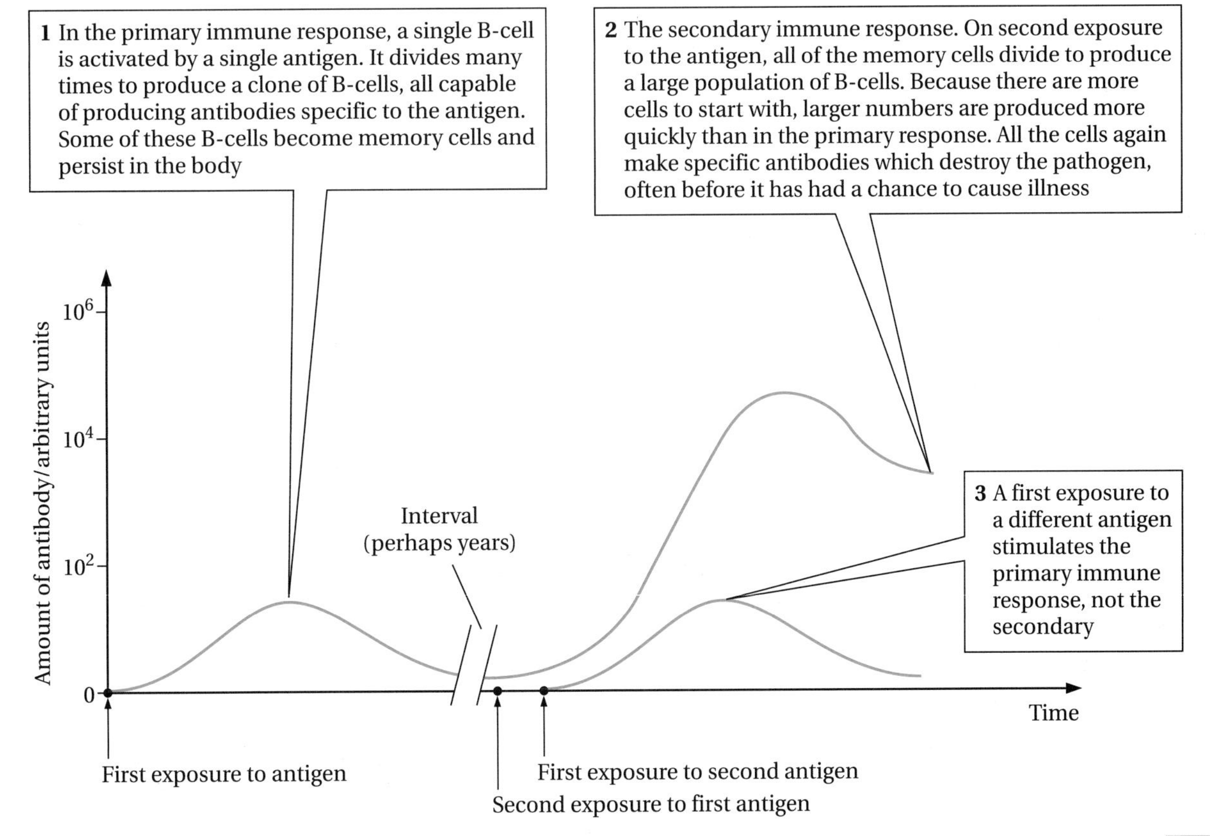

Passive and active immunity

In **active immunity** individuals make their own antibodies.

In **passive immunity** individuals get their antibodies ready made, usually from their mother.

The big problem with the immune response is that, as stated above, an individual is often not able to prevent an infection unless it has encountered the pathogen before. This would seem to leave young babies very vulnerable, and likely to die from the first infection they encounter. Help comes from the mother, who gives her baby antibodies in two ways:

- across the placenta before birth
- in her milk after birth – antibodies are proteins and you might think that they would be digested, but the baby's gut is adapted to absorb the antibodies unchanged.

Note that passive immunity can be given in another way – by an **antiserum** that contains particular antibodies. For instance, snake bites can be treated with antisera containing antibodies that neutralise the snake toxin. This is not the same as a vaccine, which does not contain antibodies. Antisera provide instant antibodies to cope with a crisis. In contrast, vaccines work in the long term, enabling an individual to *make their own* antibodies.

D

An antiserum is a sample of serum that contains particular antibodies.

A *vaccine* is the actual fluid that is injected or swallowed. *Vaccination* is the process of administering the vaccine.
The terms *vaccination* and *immunisation* basically refer to the same process.

Vaccinations

Humans have survived for thousands of years, with medical help being a relatively recent development. Until the 20th century, infant mortality was very high throughout the world. A large proportion of children – up to 50% or more – did not survive until their fifth birthday because of a combination of infectious disease and malnutrition. A major step towards reducing infant mortality in developed countries was the development of **vaccines.**

Vaccines usually contain either the pathogen or the antigens, treated so that they cannot cause the disease. The idea is to stimulate the primary response, so that when the actual pathogen is encountered, the secondary response is strong enough to prevent the disease developing.

There are several different types of vaccine:

- live, attenuated vaccines – these contain pathogens that have been treated in some way so that they can divide a few times in the body but cannot set up an infection. Measles, mumps and rubella can all be prevented by live vaccines

- dead microorganisms – these obviously cannot cause the disease but they contain the antigens that stimulate the immune response, e.g. diphtheria
- purified antigens – often made by genetic engineering (see page 33) e.g. hepatitis B.

12.4 Genetic information is passed from cell to cell during division

There are two ways in which a cell can divide: **mitosis** or **meiosis**.

D

***Mitosis** is cell division in which the DNA replicates and each daughter cell receives an exact copy of the original DNA, unless there is a mutation.*

Organisms grow and repair themselves by mitosis.

D

***Meiosis** is a more complex type of cell division with two essential features:*

- *the chromosome number is halved*
- *the genes are 'shuffled' so that each cell contains different combinations of **alleles**.*

In humans and all other animals, the gametes (sex cells: eggs and sperm) are made by meiosis.

Mitosis

When DNA is spread out in the nucleus, it is known as **chromatin**.

Mitosis is cell division that produces two daughter cells with identical genetic information. To do this the cell must first duplicate all of its DNA, and then organise the division so that each new cell gets a full set. DNA replication (copying) takes place before cell division, while the DNA is spread out rather than tightly coiled as it is in the chromosome.

Chromosomes are only visible during cell division.

Early in cell division the DNA condenses into **chromosomes**.

When chromosomes condense, they appear as double structures (Fig 13). This is a consequence of DNA replication. Each part of the chromosome – known as a **chromatid** – is identical. The two chromatids are held together by a **centromere**.

The **spindle** is the name given to the framework of protein fibres that develops during prophase. The spindle fibres are connected to the chromatids at the centromere (Fig 13). The proteins in the spindle fibres can slide over each other, bringing about movement of the chromosomes during cell division.

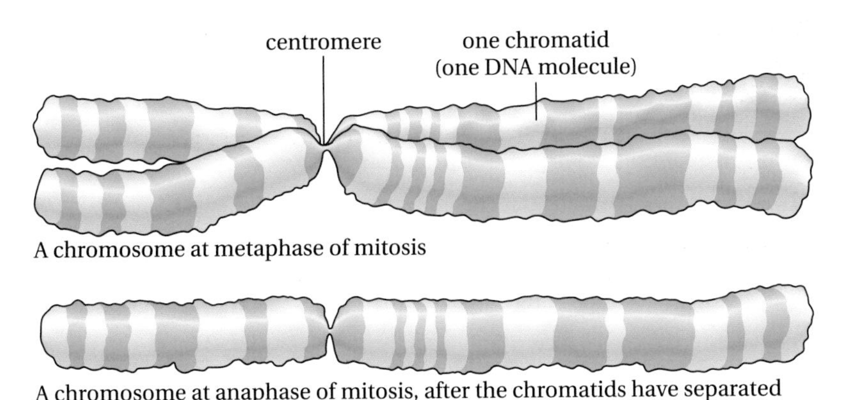

A chromosome at metaphase of mitosis

A chromosome at anaphase of mitosis, after the chromatids have separated

Fig 13
At the start of cell division, chromosomes appear as double structures consisting of two identical chromatids joined at the centromere. At anaphase, the chromatids are pulled apart, so each chromatid becomes a single chromosome

Fig 14
The stages of mitosis

> The stages of mitosis can be remembered with the mnemonic IPMAT.
> **E**

Early prophase
chromatin thread
nuclear membrane
cytoplasm
cell surface membrane
centrioles

Prophase is the first obvious stage of cell division. The key events are:

- the nuclear envelope begins to disappear
- the spindle develops
- chromosomes condense and become visible

Late prophase
nuclear membrane
centriole
centromere
pair of chromatids

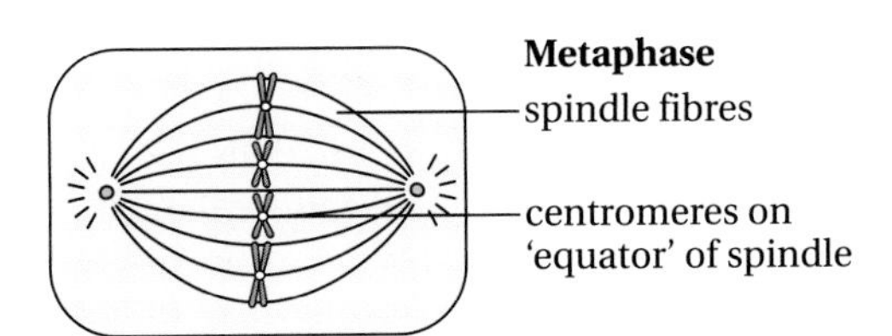

During **metaphase** the nuclear membrane disappears. The chromatids line up along the equator (middle) of the spindle

> Remembering that meta = middle and that ana = apart makes the stages easier to identify.
>
> When interpreting diagrams or photographs, remember to look at the number of strands: two long strands = prophase; two short strands = metaphase; single strands near to the equator = anaphase; single strands near to the poles = telophase.
> **E**

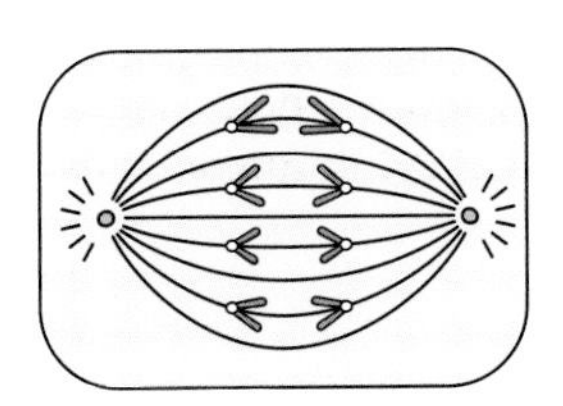

Anaphase
Daughter chromosomes move apart, led by their centromeres

In **anaphase** the chromatids are pulled apart by the spindle fibres and become chromosomes. The two sets of single chromosomes are pulled to the two poles

Telophase
nuclear membrane
chromatin threads
pair of centrioles

In **telophase** a nuclear membrane forms around each set of chromosomes. The cytoplasm constricts in the middle, so that the cytoplasm divides and two new cells are formed

Immediately after mitosis the daughter cells are small, with a relatively large nucleus. As the cell matures the volume of cytoplasm increases until the cell reaches full size.

Staining techniques – the root tip squash

This is one of the practical techniques required by the specification. The basic idea is to find tissue that is rapidly dividing so that you can see mitosis in action. One of the best is the root tip of an onion, which is easy to grow and has few chromosomes.

The technique is as follows:

1 cut off about 5 mm from the root tip – the main area of cell division is just behind the tip

2 put the tip into **acetic orcein** stain with a few drops of hydrochloric acid. The orcein stains DNA, while the acid makes the cells more permeable, allowing the stain to penetrate faster

3 warm (do not boil) the mixture on a watch glass for about 5 minutes. This also speeds up the staining process and intensifies the stain

4 place the stained tip on a clean microscope slide

5 add a few more drops of acetic orcein and gently flatten the root tip with the side of a needle

6 finally, squash the tip gently with a coverslip and observe under low and high power. Squashing the root will produce a layer that is one cell thick. Low power will allow you to isolate cells that are dividing, and the high power will allow you see the different stages of mitosis.

The cell cycle

The **cell cycle** is the name given to the whole sequence of events in the life of an individual cell. Unless the cell dies, the cycle starts and ends with cell division. In both meiosis and mitosis, the DNA must replicate *before* cell division can take place.

There are many cells in the human body that do not divide at all, e.g. adult brain cells. Cells like this remain permanently in interphase and there is no need to replicate the DNA. Other cells – e.g. skin cells – divide rapidly and so interphase is relatively short. If a cell is going to divide, the DNA will replicate *in interphase* shortly before cell division takes place.

Meiosis

Meiosis is a special type of cell division that halves the chromosome number of a cell. For this reason it is known as a **reduction division**.

There are two key features of meiosis:

- it halves the chromosome number
- it produces variation so that no two daughter cells are the same; this is why children born to the same parents are never the same unless they are identical twins (details of meiosis can be found in A2 Module 5).

In most species, but not all, meiosis produces the **gametes**, or sex cells. In humans, meiosis takes place in the ovaries and testes where diploid cells undergo meiosis to become haploid. The diploid number is restored at fertilisation (Fig 15).

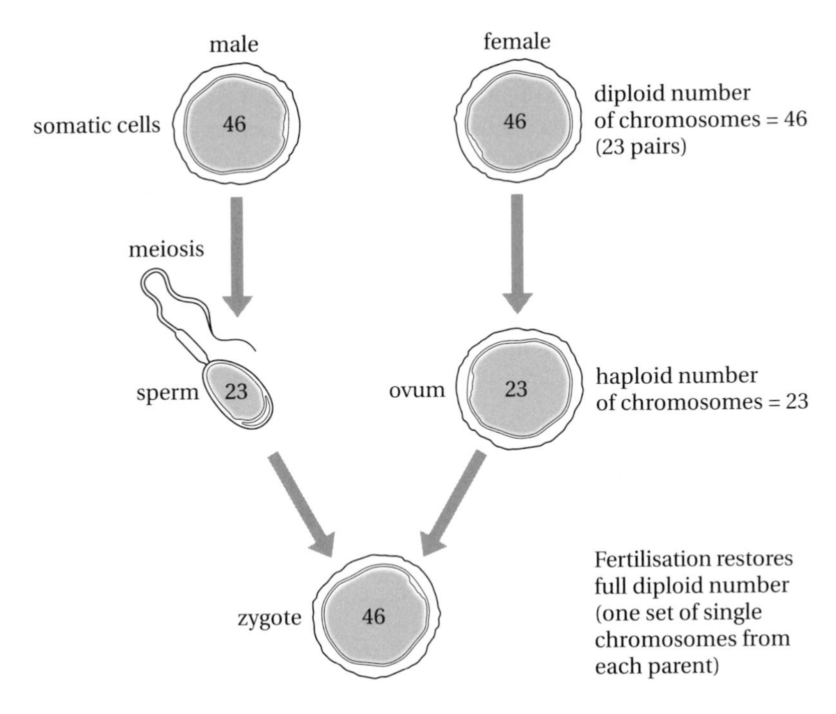

Fig 15
Most human cells contain 23 pairs of chromosomes – 46 chromosomes in total. Meiosis splits up the pairs so that each daughter cell has just one member of each pair, and 23 chromosomes in total. The diploid number is restored at fertilisation

Meiosis has a central role to play in **sexual reproduction**, a process that usually involves mixing the genes of two genetically different individuals and so producing **variation**. The importance of variation is seen in evolution; if there is a lot of genetic variation in a population, there is a greater chance that some individuals will be able to survive, especially in changing or unfavourable conditions. If there is no variation, there is nothing for natural selection to act upon – no individuals will have unique advantages that can be passed on to the next generation.

12.5 Genes incorporate coded information which determines the metabolism of organisms

DNA as genetic material

How do we know that DNA is the substance that allows characteristics to be passed on from one generation to the next? Some key experiments and breakthroughs are listed in Table 3. However, for examination purposes you do not need to memorise the list, it is more important to be able to examine and interpret evidence.

Table 3

1928	Fred Griffiths	Proposed that some 'transforming principle' changed a harmless strain of bacteria into a lethal one – the hunt for DNA began
1941	Beadle and Tatum	Irradiated the bread mould *Neurospora*, producing mutations which suggested that genes code for enzymes
1944	Oswald Avery	Purified the 'transforming principle' in Griffiths' experiment, showing it to be nucleic acid (DNA)
1950	Erwin Chargaff	Discovered that the base-pairing ratios in DNA were always the same, whatever the organism (Chargaff's principle: A = T, C = G)
1951	Rosalind Franklin	Obtained high-quality X-ray diffraction studies of DNA, showing that it had a helical structure
1952	Hershey and Chase	Used bacteriophages to show that DNA, not protein, is the material of heredity
1953	Crick and Watson	Built on the work of Chargaff and Franklin to work out the three-dimensional structure of DNA
1958	Meselson and Stahl	Used radioisotopes of nitrogen to prove the semi-conservative mechanism of DNA replication

The structure of nucleic acids

DNA

DNA stands for **deoxyribonucleic acid**. It has two key abilities:

- it carries information – the **genetic code** – from which essential proteins are made
- it can make exact copies of itself time and time again.

Genes are sections of **DNA** that code for the manufacture of particular proteins. **Protein synthesis** is covered on page 28. DNA copying, or **replication**, is covered on page 26.

DNA is a polymer: the monomers are **nucleotides**.

DNA is a **nucleic acid**, so called because it is found in the nucleus (of eukaryote cells) and is weakly acidic. It is a **polynucleotide**, i.e. made from many nucleotide units (Fig 16) joined in a chain.

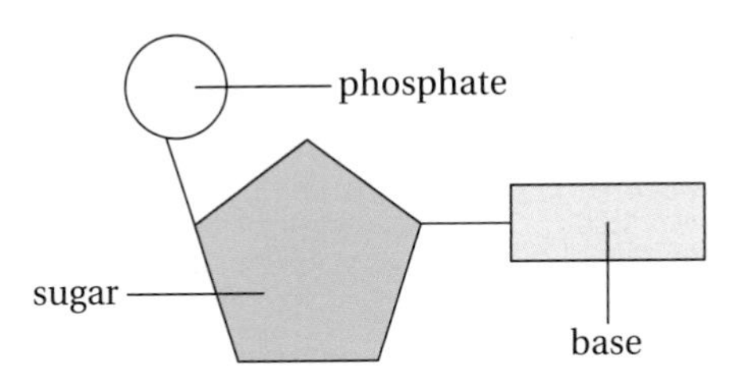

Fig 16
The basic structure of a nucleotide

Each nucleotide has three components:

- a **sugar** – deoxyribose – which is a 5-carbon sugar
- a **phosphate** group
- a **base** – one of four nitrogen-containing compounds: adenine (A), thymine (T), guanine (G), cytosine (C)

The nucleotides are arranged in a double helix, like a twisted ladder (Fig 17). The two 'sides' of the ladder are chains of alternating sugar–phosphate groups, while the 'rungs' are made from pairs of bases joined together by hydrogen bonds. It is important, both for protein synthesis and replication, that the strands can separate and rejoin without damaging the molecule. Only one part of one strand of the DNA at any particular point in the double-stranded molecule is used to make proteins – this is called the **sense strand**. The other side serves to stabilise the molecule. The sense strand for different genes may be found on different sides of the molecule.

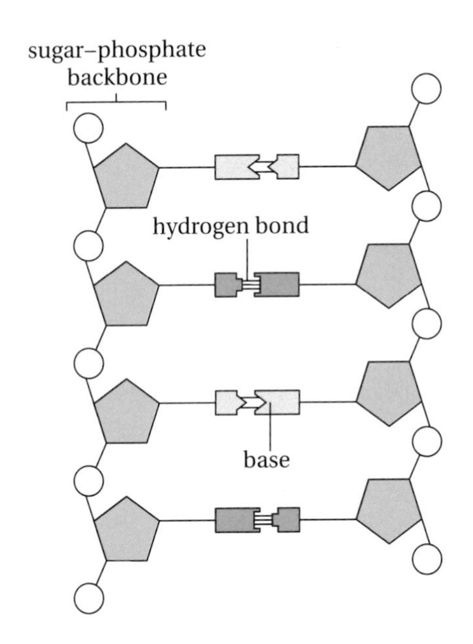

Fig 17
Nucleotides are joined by condensation reactions to form a single DNA chain which bonds with a complementary DNA chain to make a double helix

The bases

The four bases always bond in the same way – A to T and C to G. So if you know the base sequence down one side of the DNA molecule, you can predict the other.

For example,

if one strand reads CGGTACCTA

the other side will read GCCATGGAT

Practise doing these – there will probably be examples in the exam.

E

The bases are held together by hydrogen bonds, two between A and T (i.e. A=T) and three between C and G (G≡C). These regular hydrogen bonds along the whole length of the molecule make DNA very **stable**. When heated, the two strands of DNA do not come apart until about 86 °C. This makes DNA much more heat resistant than proteins, many of which are denatured at about 50 °C.

RNA

RNA stands for **ribonucleic acid**. There are several types of RNA, which are also nucleic acids like DNA, although they are not always found in the nucleus. There are two main types:

- **messenger RNA** (mRNA) is a single long strand of nucleotides that is a copy of a gene (Fig 18)
- **transfer RNA** (tRNA) is a small, cloverleaf-shaped molecule that brings particular amino acids to the ribosome during translation.

Fig 18
The structure of messenger RNA

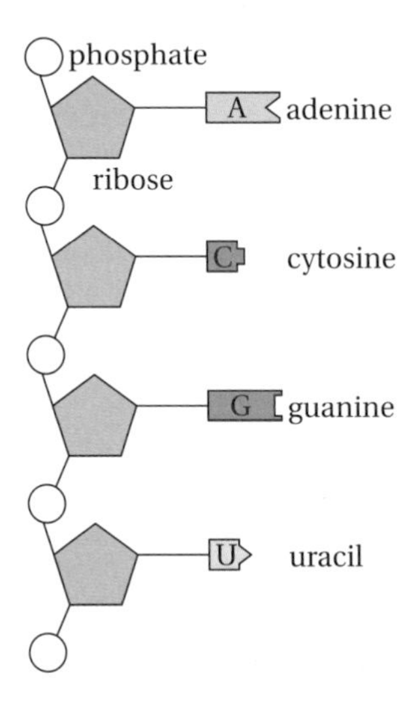

The essential differences between DNA, mRNA and tRNA are summarised in Table 4.

Table 4
A comparison of DNA, mRNA and tRNA

	DNA	**mRNA**	**tRNA**
sugar	deoxyribose	ribose	ribose
bases	C, G, A and T	C, G, A and U	C, G, A and U
strands	double	single	single
lifespan	long term	short term	short term
site of action	nucleus	nucleus and cytoplasm	cytoplasm

Replication of DNA

The existence of a molecule that can store information and copy itself is essential to life. Otherwise, the characteristics of life cannot be passed on from generation to generation.

In DNA replication the two strands come apart, each one acting as a template for the addition of matching nucleotides (Fig 19). Enzymes separate the two strands, and DNA binding proteins keep the strands apart. DNA polymerase enzymes then move along the exposed strands, catalysing the addition of nucleotides to complete the new strands.

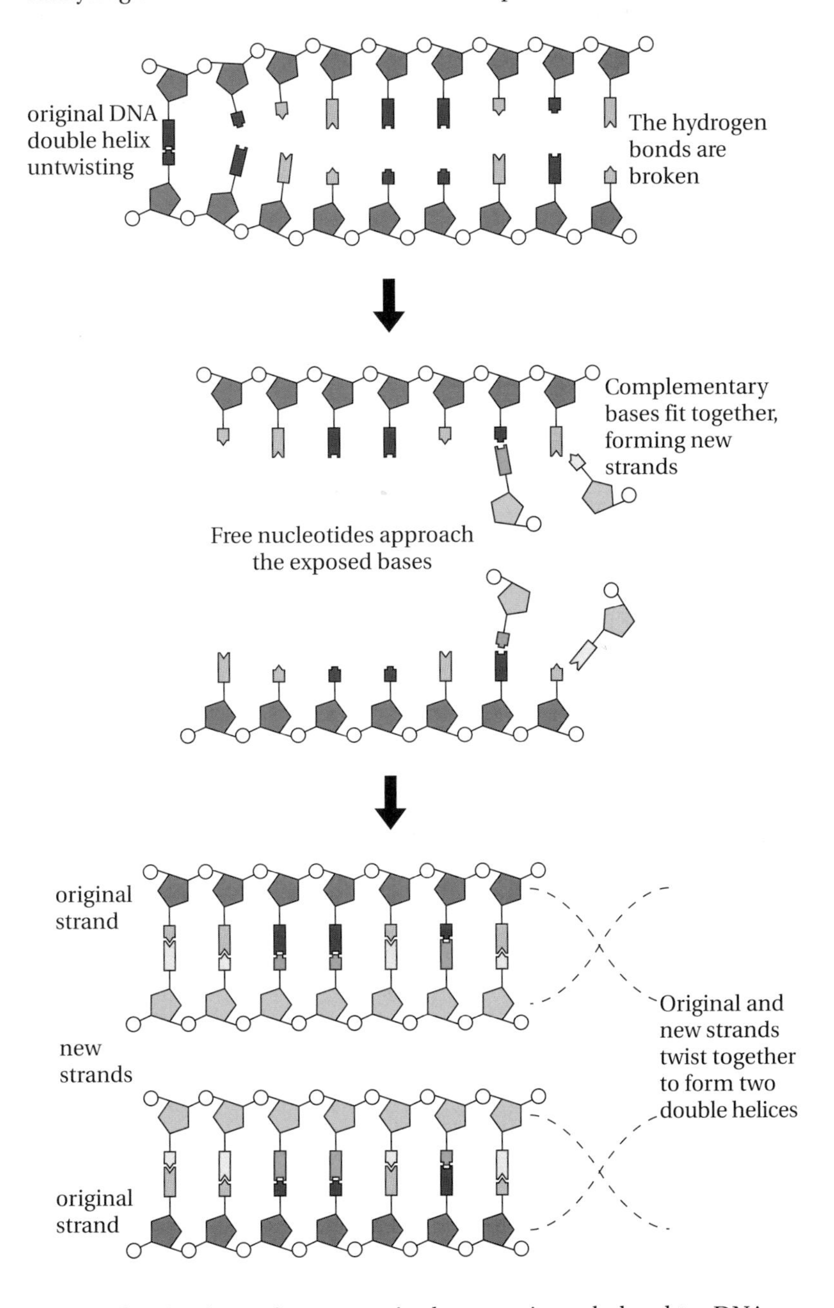

Fig 19
DNA replication

DNA replication is **semi-conservative** because, in each daughter DNA molecule, one strand is original (i.e. has been conserved) and the other strand is new.

Proof that DNA replication is semi-conservative came from a series of classic experiments performed by Meselson and Stahl in the 1950s. They used an isotope of nitrogen, ^{15}N (heavy nitrogen), which can be incorporated into DNA instead of the normal isotope, ^{14}N, without harming the organism. However, this makes the DNA slightly denser and so DNA containing ^{14}N can be separated from that containing ^{15}N by centrifugation (Fig 20). In this way they were able to prove that each new DNA molecule contains one original strand and one new strand.

Fig 20
An outline of Meselson and Stahl's experiment

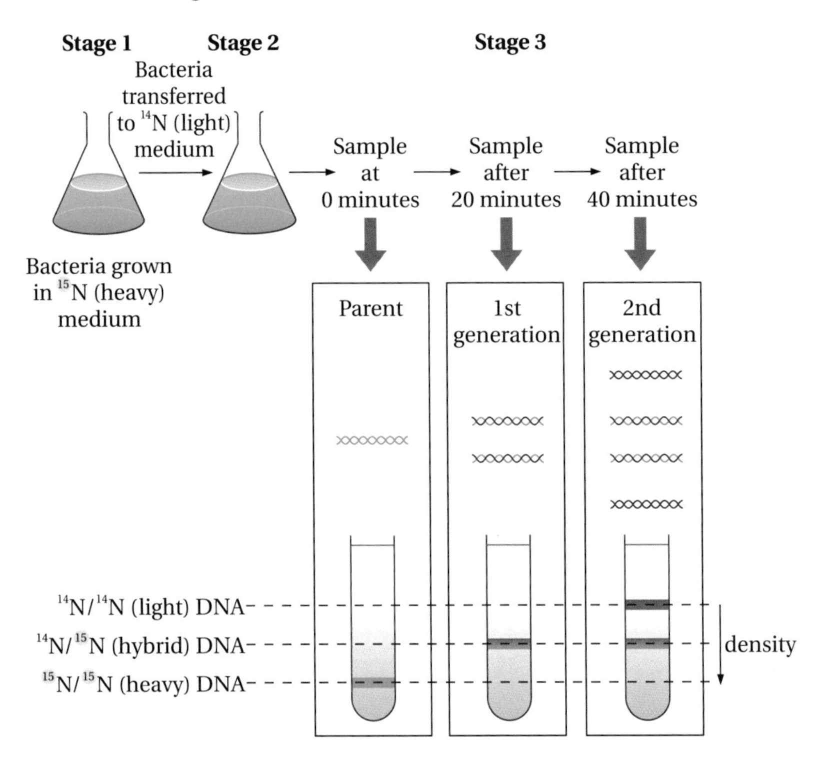

E Exam questions often describe Meselson and Stahl's experiment and ask candidates to predict the bands obtained in the second and third generation.

- generation 0 = DNA all 'heavy' (both strands heavy)
- first generation = two hybrid molecules (all DNA molecules have one light strand and one heavy)
- second generation = two hybrid molecules and two all new (light)
- third generation = two hybrid molecules and six all light, etc.

Remember that in each generation there will be the original number of 'heavy' strands. All of the new ones will be made from 'light' DNA. With each new generation the originals get more and more swamped by the new molecules.

Protein synthesis

The genetic code

The genetic code is the sequence of bases in DNA that codes for the order in which amino acids are assembled in a polypeptide or protein.

There are four bases that code for 20 different amino acids, so:

- one base cannot code for one amino acid
- a two-base code would give $4 \times 4 = 16$ combinations, which is still not enough
- a three-base code gives $4 \times 4 \times 4 = 64$ possible different combinations, which is more than enough.

D ***A sequence of three bases is known as a codon.***

One codon codes for one amino acid, e.g. the codon AAA codes for the amino acid phenylalanine. As there are 64 codons, some amino acids are coded for by more than one codon. Some also act as a 'full stop' to stop the amino acid chain growing.

The genetic code can be described as a **non-overlapping, degenerate code**.

- *Non-overlapping* indicates that the *codons* do not overlap. For example, the DNA sequence CTACTA is two codons (CTA, CTA), although there would be more if the codons overlapped (CTA, TAC, ACT, CTA)
- *Degenerate* indicates that there are 64 different codons but only 20 amino acids, so there can be *more than one codon* for any particular amino acid.

E

You will not be expected to know any examples of the genetic code. Exam questions will always give the information needed to match any relevant codons to particular amino acids.

D

***A gene** is a length of DNA that contains all the codons needed to synthesise a particular polypeptide or protein.*

The hormone insulin, for example, consists of 51 amino acids and so the insulin gene has at least 51 codons, or 153 bases.

Role of nucleic acids in protein and enzyme synthesis

How is the genetic code in DNA used to build a protein? There are two stages:

- transcription
- translation.

***Transcription** is when the DNA code on a gene is copied by another molecule, **messenger RNA** (mRNA).*

The mRNA molecules are effectively mobile copies of genes. They carry the code out of the nucleus to the site of translation, on the **ribosomes** in the cytoplasm.

***Translation** is when the genetic code in the mRNA molecule is used to build the polypeptide or protein.*

The process of transcription

Most cells in the human body contain two complete sets of genes. However, only a small proportion of genes are used – or **expressed** – in any particular cell. For example, every cell in the human body contains two copies of the gene that codes for insulin, but only certain specialised cells in the pancreas actually use the gene to make insulin.

Step 1 – The two strands of DNA unwind along the length of the gene (Fig 21). This is catalysed by **enzymes**.

Step 2 – The enzyme **RNA polymerase** moves along one side of the DNA molecule – the **sense strand** – that contains the genetic code. The enzyme catalyses the assembly of an mRNA molecule by the addition of matching nucleotides. When mRNA is synthesised, the base **thymine** is replaced by **uracil**, so the base pairing between DNA and mRNA is always A with U and C with G.

Step 3 – The mRNA molecule peels off the gene and passes out of the nucleus.

Fig 21
The steps in transcription

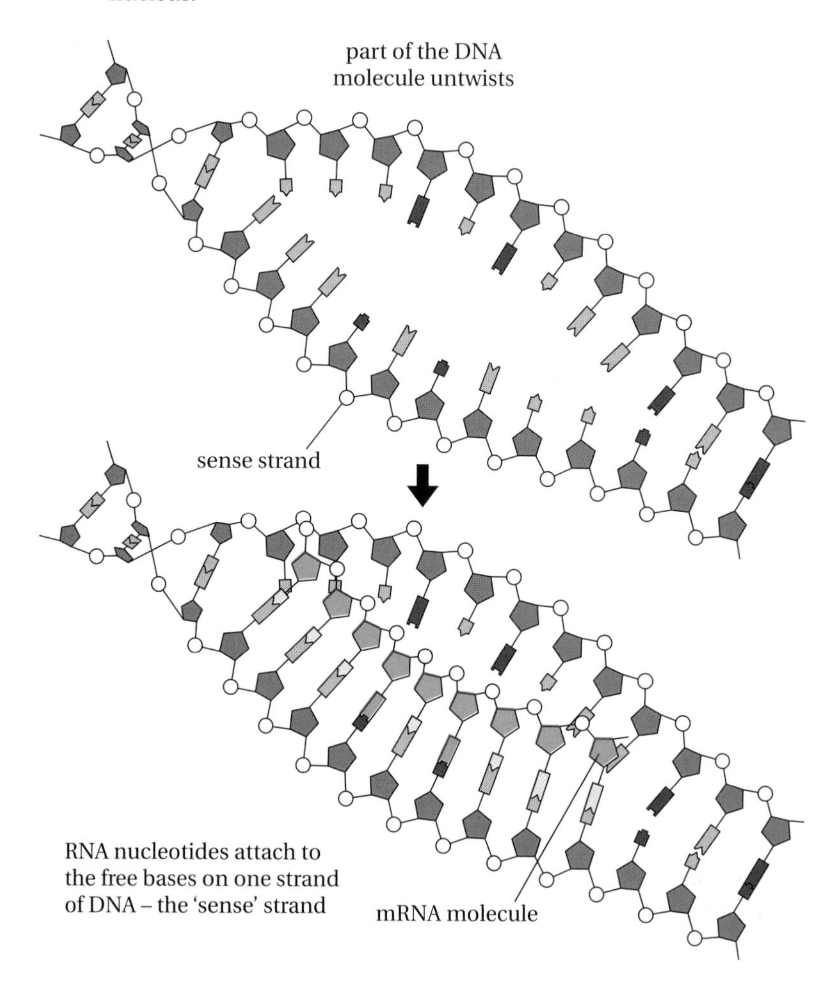

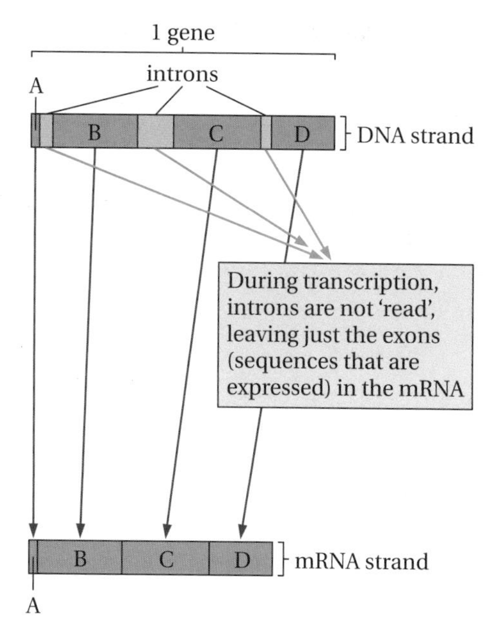

Fig 22
A gene being transcribed, to omit the introns

Strangely, the codons in a gene are not consecutive – there can be gaps in between the codons consisting of base sequences that code for nothing at all. For instance, the protein insulin consists of 51 amino acids and so you might expect the insulin gene to consist of 153 bases. However, it contains more because the coding regions are separated by several non-coding regions. The non-coding base sequences are called **introns**, while the coding parts are called **exons**. During transcription, the introns are omitted, so the messenger RNA consists of the exons joined together (see Fig 22). Remember, the **ex**ons are the sequences that are **ex**pressed, i.e. used to make a protein.

Translation

Transfer RNA transfers amino acids to ribosomes during translation. At one end of the molecule is the **anticodon**, which is a three-base sequence that binds to the codon on the mRNA. At the other end is the amino acid specified by the codon (Fig 23). For example, the codon AAA will have an anticodon of UUU.

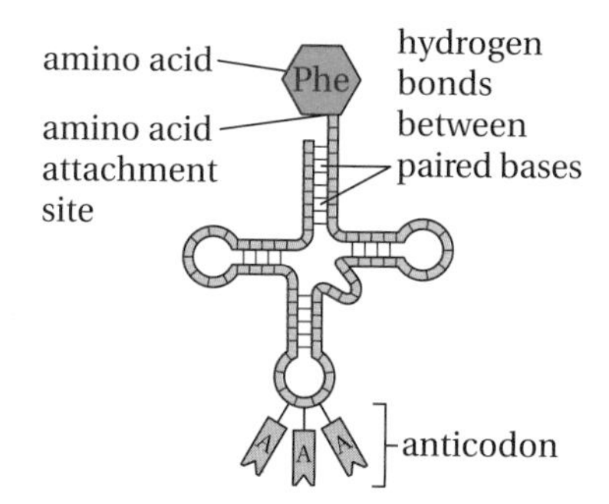

Fig 23
The structure of transfer RNA

A **ribosome** is a small organelle that can be thought of as a giant enzyme. Ribosomes hold together all the components of translation (Fig 24):

Step 1 – The mRNA attaches itself to a ribosome.

Step 2 – The first codon is translated. For instance, if the codon reads UUU, which codes for the amino acid phenylalanine, a transfer RNA molecule with the anticodon AAA attaches, carrying a phenylalanine molecule at the other end.

Step 3 – The second codon is translated in the same way. The second amino acid is held alongside the first, and a peptide bond is formed by condensation between them. The polypeptide chain has started.

Fig 24
The overall process of translation

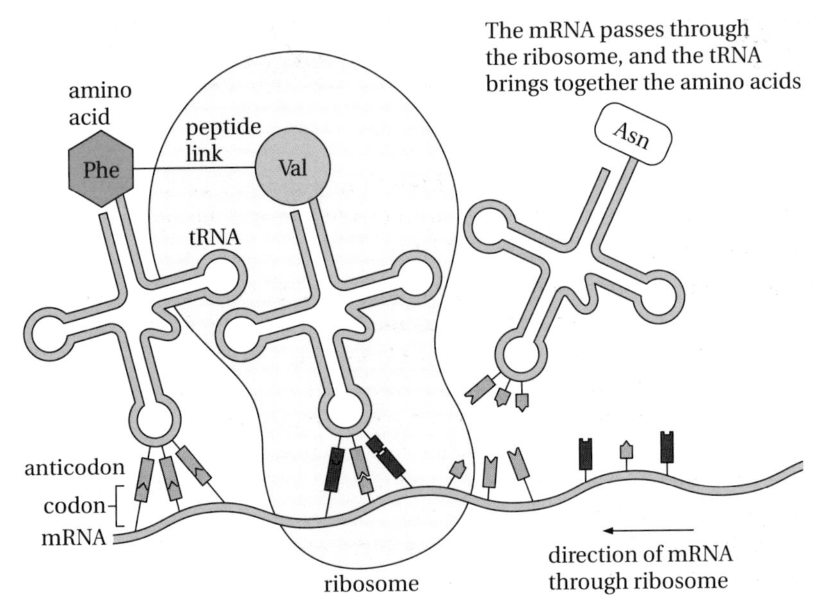

Step 4 – The process is repeated, the mRNA moving along the ribosome until the polypeptide has been built. If a stop codon is encountered, translation ceases and the polypeptide is finished.

E The problems you will be given usually involve both transcription and translation, and will test whether you realise that A in DNA is transcribed into U in mRNA and that the corresponding base in tRNA is A. Make sure that you can complete the table below

Base in DNA	C	G	A	T
Base in RNA				
Base in tRNA				

Protein synthesis and disease

In genetic disease, the problem is often that a gene has mutated (changed) so that it no longer codes for the required protein. When the faulty code is translated in protein synthesis, the protein made is the wrong shape and will not function properly. So, usually, the symptoms of disease are the result of some vital protein not doing its job.

In **cystic fibrosis**, the problem is that one gene has mutated so that it no longer codes for a vital membrane protein. This protein transports ions out of epithelial cells, particularly those lining the lungs. In healthy people, the protein pumps chloride and sodium ions out into the mucous lining of the lungs. The ions draw water out of the cells by osmosis, making normal, watery mucus that can be moved along by the tiny hairs (cilia) that line the lungs. If there is no protein, the cell cannot pump ions and the mucus becomes very thick, so that it cannot be dislodged by the cilia. This makes an ideal breeding ground for bacteria, so lung infections are a major problem.

Cystic fibrosis is treated by patting the chest to dislodge the mucus, and taking antibiotics to fight infection. **Gene therapy** trials, in which healthy copies of the cystic fibrosis gene are inhaled and are taken in by the epithelial cells, are under way.

12.6 Gene technology may be used in combating disease

Recombinant DNA

In recent years we have progressed from simply studying patterns of inheritance to locating genes, finding out what they do, cutting them out and inserting them into other organisms.

A classic application of genetic engineering is the production of **human insulin**. There is a huge demand for insulin to treat diabetics, but this complex protein cannot be made in the laboratory, and it cannot be extracted from human pancreases because there aren't enough available. Until recently, pork or beef insulin was used, but this is not the same as human insulin and becomes less effective with time. However, once the gene for human insulin was found, it could be cut out, cloned and placed into bacteria. When cultured, the bacteria made large amounts of valuable human insulin. Many other valuable proteins, such as factor VIII for haemophiliacs, are also made this way.

D

***Genetic engineering** is the transfer of genes from one organism to another (often from one species to another).*

*A **genome** is the entire DNA sequence of an organism; the genes and all the non-coding DNA in between.*

The genome of some simple organisms was worked out some years ago, e.g. the bacterium *Escherichia coli* (*E. coli*), followed by the fruit fly *Drosophila.* Now the entire human genome has been mapped – about 60 000 genes and 3 billion base pairs.

D

***Transgenic organisms** are organisms that have had their DNA altered by the insertion of genes from another organism.*

For example, some cereal crops have had genes for disease resistance inserted. Most of our genetically modified foods have had beneficial genes inserted into their genome – a source of heated debate.

D

***Recombinant DNA** is DNA that has been cut and re-joined in a different way. Often, the DNA of one species has one or more genes from another species inserted into it.*

Transgenic organisms therefore contain recombinant DNA.

Isolating the gene

There are several different approaches to finding a particular gene:

- make a **genetic probe** – if you know part of the base sequence of the gene you want to find, you can make a DNA probe. This is a short piece of DNA of bases that are complementary to the known sequence, with a radioactive or fluorescent label. The genetic probe will seek out and bind to the gene, indicating its position
- use messenger RNA – if you can isolate the mRNA that is a copy of the gene you want to find, you can use it as a DNA probe as described above
- work backwards from the protein – if you know the amino acid sequence of the protein that the gene codes for, you can make a piece of DNA that codes for this amino acid sequence. You can then insert this artificial gene into an organism.

Three essential types of enzyme used in genetic engineering are:

- **restriction endonucleases** cut DNA
- **DNA ligases** stick DNA together again
- **reverse transcriptases** make DNA from RNA.

Enzymes – tools of the trade

Restriction endonucleases – called **restriction enzymes** for short – cut DNA strands at precise points, known as **recognition sites**. The recognition site is a particular sequence of bases. For example, the enzyme *Eco*R1 (extracted from the bacterium *E. coli*) has a recognition site of GAATTC and so will cut DNA whenever it comes across that sequence (Fig 25). These enzymes produce staggered cuts known as **sticky ends**.

There are many different restriction enzymes, each with a unique recognition site. This enables genetic engineers to select a particular enzyme for each job, so that cuts can be made where required. This enables us to cut out different genes with precision.

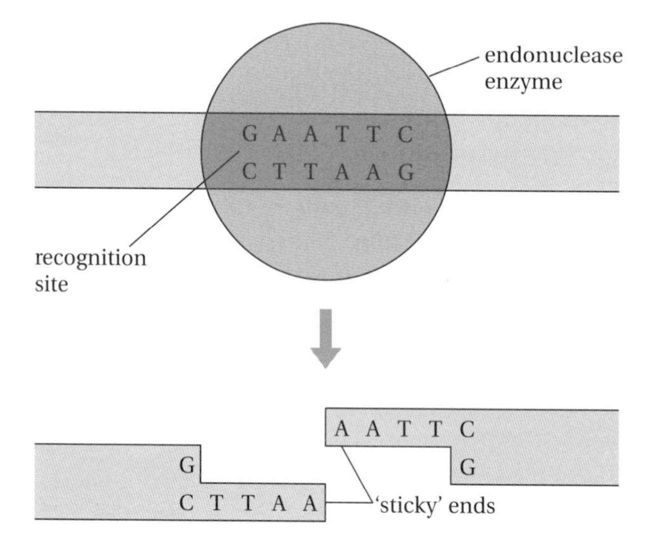

Fig 25
Most restriction enzymes make staggered cuts, rather than clean ones, so that a few bases are exposed – four in this example. This allows it to be connected to any other piece of DNA with a matching base sequence, such as one that has been cut with the same enzyme

Once the DNA has been cut out it can be joined – **spliced** – into a different piece of DNA. If the same restriction enzyme is used to cut both pieces of DNA, complementary sticky ends will be produced. A **DNA ligase enzyme** will then join the sticky ends together.

Reverse transcriptase enzymes can be used to make DNA from mRNA – transcription in reverse. This enzyme is found in retroviruses such as HIV (see page 6).

Genetically modified microorganisms

Once the insulin gene has been isolated, cut out and copied, it needs to be inserted into a bacterium. This is often achieved using **plasmids**, which are small circles of DNA that are found in bacterial cytoplasm. Plasmids are easy to extract and are commercially available. The gene can be spliced into the plasmid using the enzymes described above (see Fig 26). The plasmids are then inserted into the bacteria. The plasmids are carriers, or **vectors**, taking the insulin gene into the bacteria. **Viruses** might also prove to be effective vectors when putting genes into eukaryotic cells, such as yeast or even human cells.

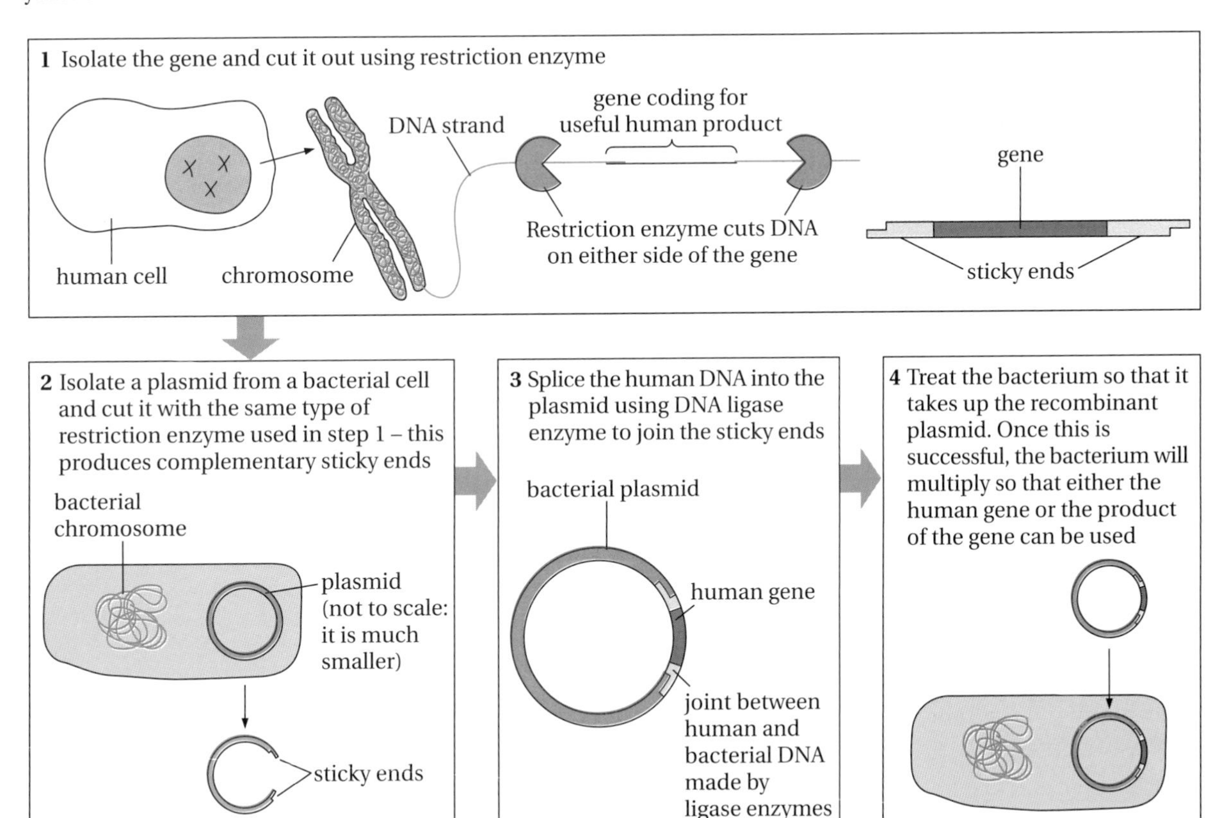

Fig 26
The basic steps in the transfer of a human gene to a bacterium

Genetic markers

The process of inserting plasmids into bacteria is very unreliable. When you mix plasmids and bacteria, for every bacterium that takes up a plasmid, thousands don't. So how do you tell which bacteria have accepted the new gene? One way is to add a **genetic marker**.

D

*A **genetic marker** is an extra gene, which is inserted into the plasmid along with the gene that is to be cloned.*

A common example of a genetic marker is a gene for antibiotic resistance. The bacteria are then grown on a medium that contains the antibiotic. Only those bacteria that took up the plasmid with the new gene will survive and grow.

Another example of a genetic marker is a gene which makes an enzyme that catalyses a reaction that forms a coloured product. The bacteria are then grown on the substrate which can be made into the coloured product. Any transgenic colonies will show up coloured and so are easily seen. These colonies can then be isolated and cultured.

Large-scale culturing

Bacteria are prokaryotes, but genes can also be put into simple eukaryotes such as yeast (a single-celled fungus). Both bacteria and yeast can be cultured on an industrial scale in large fermentation chambers. Given the right conditions (see page 4), the microorganisms will reproduce asexually, cloning themselves – and the new genes – every half hour or so. The products can then be extracted and purified for use. Examples of the products made in this way include human insulin and other human hormones, antibiotics and enzymes.

Evaluation of genetic engineering

The advances in gene technology are such that we will have the means to perform many activities which seemed science fiction even just a short time ago. Should we do something just because we can? There are people who would ban any interference with genetic material, claiming that we are 'playing God'. Certainly the implications of any activity need to be carefully evaluated. Table 5 lists some of the more recent activities and issues in genetic engineering.

E

You should be able to give advantages and disadvantages of the consequences of genetic engineering, e.g. rearing pigs so that we can transplant their organs into humans. Try to be more specific than simply saying 'there are religious difficulties' or 'we must not play God'. A little more in the way of explanation is required. Your points can include ethical, moral, social and environmental considerations, but the examiners will mainly be looking for biological answers.

Activity	Possible objection
Inserting human genes into other animals, e.g. sheep, so that the gene is expressed and the product can be purified from the milk	Unknown long-term effects of interfering with a complex organism's genome
Inserting human genes into microorganisms, culturing on an industrial scale	Who 'owns' the product – the company that makes it or the original source?
Transferring beneficial genes into plants	Possible danger when interfering with an organism's genome is that it may have unexpected effects, e.g. toxic by-products, effects on the ecosystem
Using embryos for research	When does an embryo become a human being, with its own rights?
Cloning organisms: e.g. embryos of prize cattle can be split many times to produce identical embryos which can be implanted into ordinary cattle	Cloning humans – possible psychological problems. Should a person be allowed to clone themselves for reasons of vanity?
Animal organs for human transplant	Concerns about safety, also the 'yuk' factor (instinctive revulsion). Some claim that it is 'speciesist' – it implies that humans are superior to other species, when morally we should grant all species equal rights
Treatment of ageing (by use of the telomerase enzyme)	Who gets the treatment – only the rich? If some people live to be 150, what effect will it have on the population/ family structure/economy/workforce?

Table 5
Some of the current activities and issues in gene technology

Any activity could be evaluated against the following criteria:

1 technical difficulty – is the activity too difficult (or expensive) to do reliably?

2 ethical problems – is the activity immoral?

3 potential catastrophe – something unknown could happen, can we predict all possible outcomes of the activity?

Criterion 1 for any activity will probably diminish with time as practical techniques and methods improve. Criterion 3 will also change as techniques and methods change; we should get better at prediction. However, criterion 2 will always be an issue.

12.7 Non-communicable disease includes heart disease and cancer

The biological basis of heart disease

Familiarise yourself with the structure of an artery wall from Module 1.

Cardiovascular disease is the biggest single cause of death in the UK, and accounts for over a quarter of all deaths – about 175 000 per year. The common underlying cause of cardiovascular disease is a build-up of a fatty material, called **atheroma**, in the walls of arteries (Fig 27).

Glossary of cardiovascular disease

atheroma – the fatty deposit that builds up under the endothelium (lining) of blood vessels. As the atheroma gets thicker, the lumen of the artery gets smaller

aneurysm – a ballooning of an artery due to a weakness in the wall. This requires surgery because a ruptured aneurysm is usually fatal.

angina – chest pains causes by an inadequate supply of blood (and therefore oxygen) to the heart muscle

thrombosis – a blood clot lodged in a vessel

embolism – a clot/thrombus travelling in the bloodstream

atherosclerosis – a build-up of atheroma in the blood vessels

arteriosclerosis – hardening of the arteries. Associated with old age, the blood vessel walls, particularly of the arteries, become less elastic and more liable to rupture

myocardial infarction – a heart attack. Part of the heart muscle – the myocardium – dies (infarcts) when the blood supply to that area is blocked.

The link between atheroma and the increased risk of aneurysm and thrombosis

The build-up of atheroma is usually the result of a combination of risk factors, both environmental and genetic:

- hypertension, or high blood pressure – this can be the result of several factors: high salt intake, smoking, obesity and a genetic tendency
- smoking – there is a well-established link between smoking and atherosclerosis, although the exact mechanism is unclear
- diet – a diet high in saturated fats and cholesterol (mainly from animal products) can lead to a high concentration of low density lipoproteins (LDLs) in the blood. High concentrations of LDLs are associated with atherosclerosis.

Overall, atherosclerosis causes disease in two ways:

- by narrowing or even blocking vital blood vessels, such as the coronary arteries which supply blood to the heart muscle

- by initiating the formation of blood clots. A blood clot can be fatal if it lodges in a coronary artery (leading to a heart attack) or in the brain (leading to a stroke).

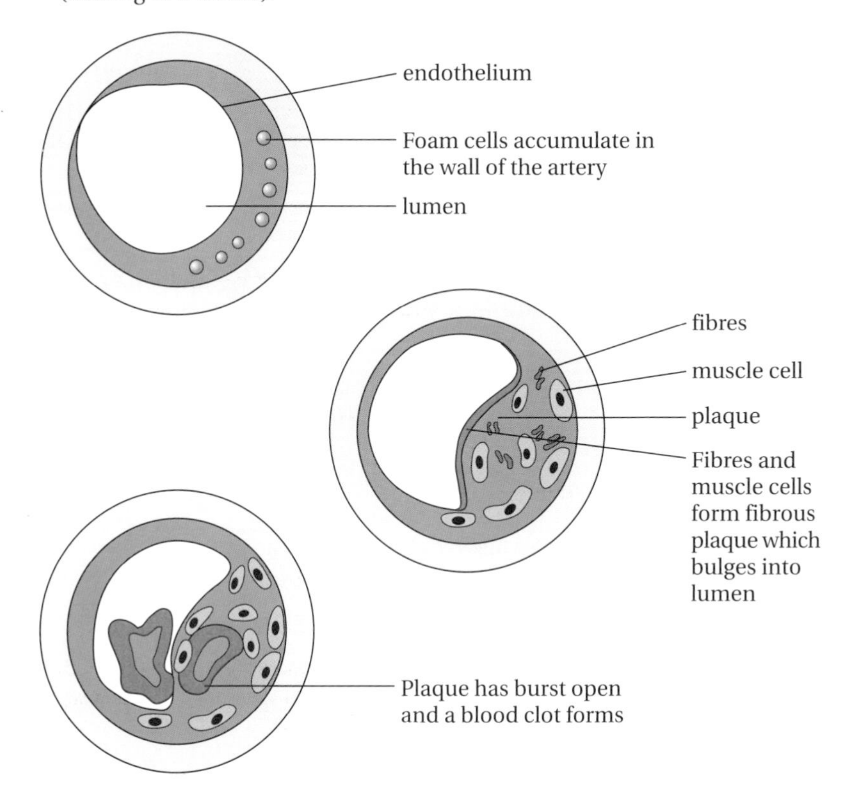

Fig 27
The build-up of atheroma in an artery. Atheroma builds up under the endothelium (lining) of the artery, causing a narrowing and roughening of the wall. This reduces blood flow and increases the chance of blood clots forming and lodging in the lumen

E

Many candidates state that atheroma is simply a build-up of fat *in the lumen* of the artery. More accurately it is a build-up of fatty material *under the endothelium* (lining) of the vessel.

The biological basis of cancer

Usually, cells in the human body only divide when they should, in order to allow growth or the repair of tissues. When the mechanisms that control cell division (mitosis) break down, the result is the uncontrolled division of cells, resulting in a tumour (Fig 28).

There are basically two types of tumour:

- **benign** – these tumours are enclosed in a capsule and grow in the centre, so they don't invade the surrounding tissues; they are not cancerous, and are often easily removed by surgery
- **malignant** – these tumours grow at the edges, invading the surrounding tissues and organs; they are cancerous, and are much more difficult to treat. Often it is difficult to tell where their boundaries are, making surgery difficult. Cells may break off and set up secondary tumours elsewhere in the body – this spreading process is called **metastasis**.

The causes of cancer

There are many factors involved in the development of cancer, including genetic and environmental factors.

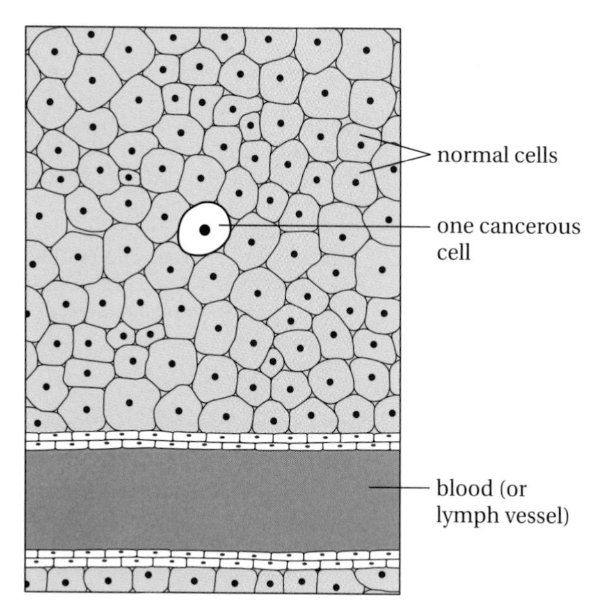

One cell starts to divide uncontrollably: it becomes cancerous

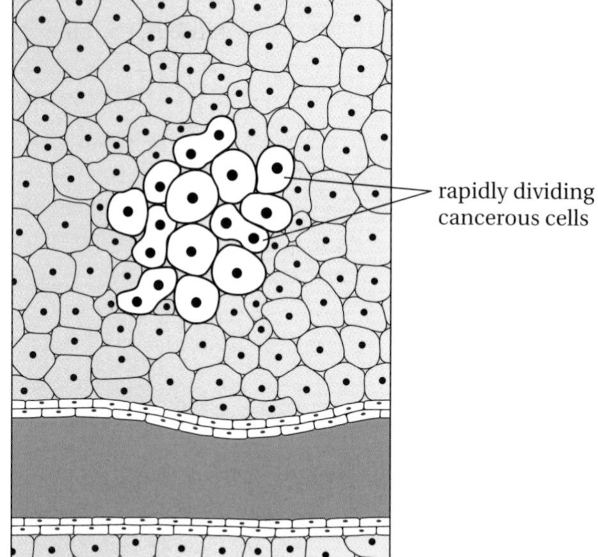

The cancerous cell divides rapidly, forming a mass of cells – the primary tumour – which squashes out the neighbouring normal cells

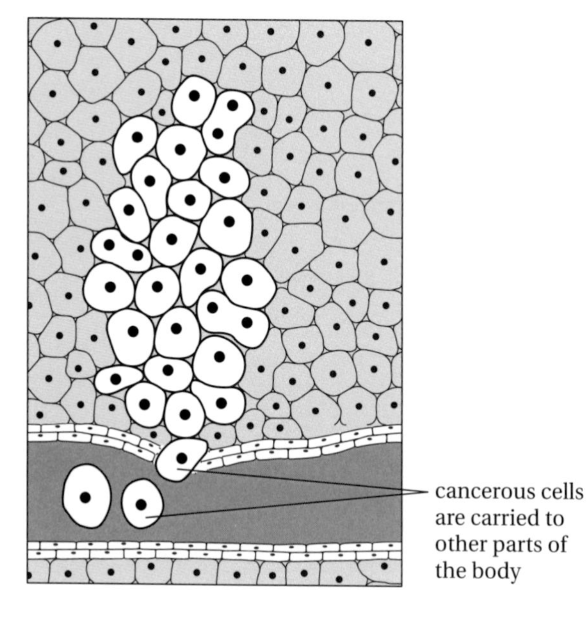

The process of metastasis. A clump of cells breaks free from the primary tumour and is then carried by the blood or lymphatic system to another part of the body. Fortunately, very few of these clumps of cells, about one in 10000, are able to establish themselves and form a secondary tumour, but that is enough. Nearly 60% of people who are diagnosed with cancer are found to have well-established secondary tumours.

Fig 28
The development of a tumour and the process of metastasis

Genetic factors

The development of cancer can be caused by a mutation of the genes that control cell division. Scientists have isolated several genes – **oncogenes** – whose mutation leads to the cell losing its ability to control cell division.

The incidence of cancer, however, is much rarer than the mutation of oncogenes. This is because there is a back-up control system in the form of **tumour suppressor genes**.

These genes prevent cells from dividing too quickly, giving time for the immune system to destroy the rogue cells, or for the damaged DNA to be repaired. If the tumour suppressor genes mutate, the cell's safety mechanisms are lost, and the development of cancer is more likely.

Cancer is commoner in older people because their somatic (body) cells accumulate mutations. Sometimes, however, these mutations occur in gametes (sex cells), so that the genes are passed on to the next generation. People who inherit these genes are said to have a **genetic predisposition** to cancer, i.e. they are more likely to develop cancer, especially at an early age.

Environmental factors

There are many factors that increase the rate of mutation of the cells described above, and so increase the risk of developing cancer. Such factors are called **carcinogens** and include:

- smoking – tobacco smoke contains a variety of carcinogens
- diet – several substances we eat or drink can cause cancer, e.g. alcohol has been linked with a higher incidence of mouth, throat and oesophageal cancer. A lack of fibre and a high intake of red meat and animal fat seems to be associated with cancer of the colon and rectum, which is very common in the western world
- radiation – certain types of radiation are known to be carcinogenic because they damage DNA. Ultra-violet radiation (in sunlight) does not penetrate far into human tissue but it can cause skin cancer. Ionising radiation, e.g. from nuclear fallout, can penetrate much further and can cause cancers such as leukaemia (cancer of the bone marrow)
- chemical carcinogens – asbestos, benzene, methanal (formaldehyde) and diesel exhaust fumes are all carcinogenic
- some microorganisms, notably viruses, have been linked with the development of cancers. The human papilloma virus is associated with over 90% of cases of cervical cancer.

12.8 Disease may be diagnosed by a variety of techniques

DNA probes and diagnosis

The development of DNA profiling techniques in the 1980s has also allowed us to test an individual's DNA to see if particular genes are present, particularly the genes coding for genetic diseases such as cystic fibrosis or haemophilia. The basic idea is that if you know part of the DNA sequence of the gene you want to test for, you can make a complementary piece of DNA – a genetic probe – that will seek out and bind to the gene. A radioactive or fluorescent marker is added to the probe so that the gene shows up.

The basic steps in the process are summarised in Fig 29.

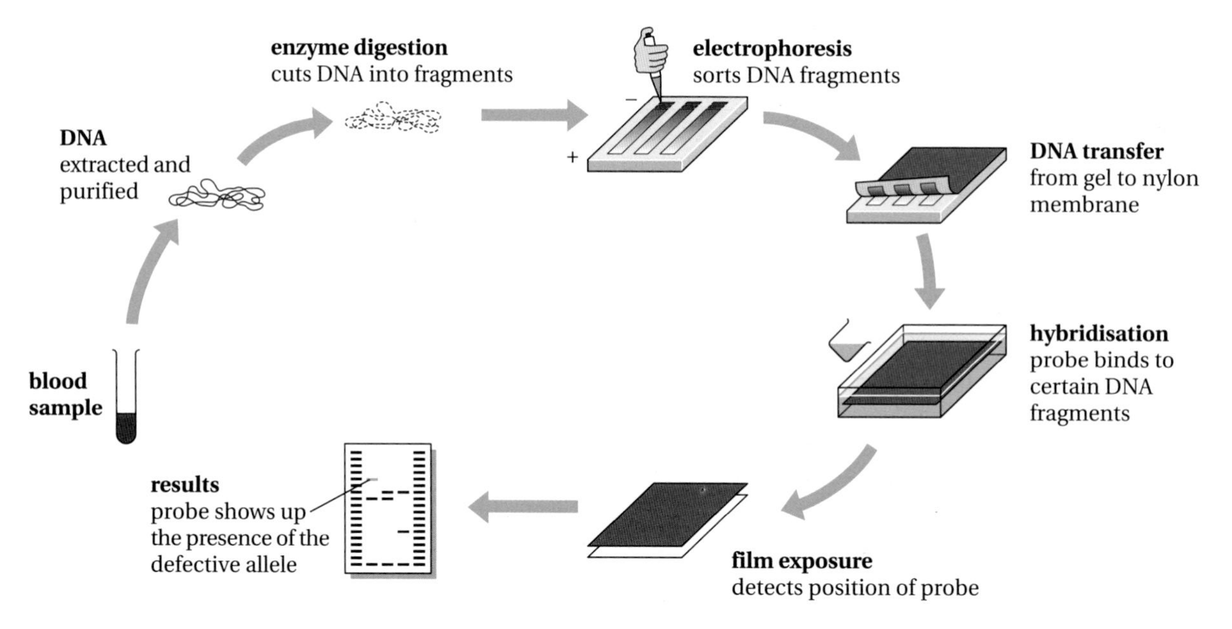

Fig 29
Using a DNA probe to detect a disease

Step 1 – The individual's DNA is extracted and purified.

Step 2 – The DNA is split into fragments by restriction endonuclease enzymes.

Step 3 – The fragments are sorted by electrophoresis, which involves applying an electric current through the mixture of DNA fragments. The short fragments of DNA move rapidly through the gel. The larger ones move more slowly.

Step 4 – Probes/markers are added that bind to the defective allele, showing its presence in an individual. The excess probe is washed off. If the individual carries the allele, it will show up, if not, the probe will have been washed away.

PCR – the polymerase chain reaction

This technique is basically DNA cloning in a test tube. Starting with a tiny sample of DNA, PCR can make millions of copies quickly.

The ingredients needed are:

- the original DNA sample
- nucleotides
- DNA polymerase enzyme
- primers.

The basic steps are shown in Fig 30.

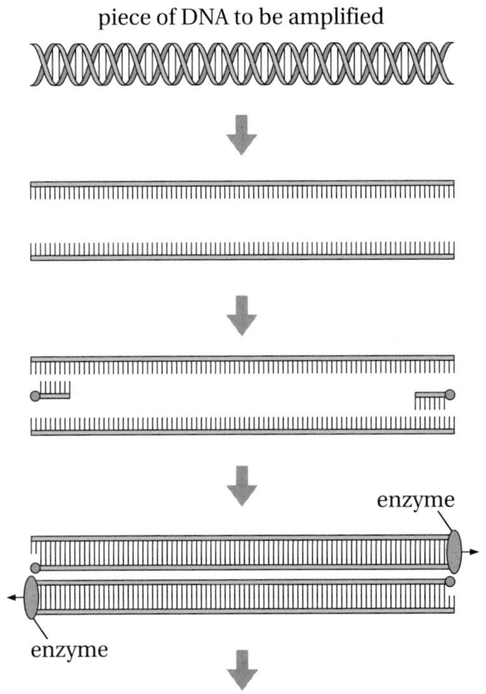

piece of DNA to be amplified

1 The sample DNA is heated so that the two strands separate

2 Primers are added. These are short lengths of DNA that effectively say 'start copying here'

enzyme

3 The DNA polymerase works its way along both strands, adding complementary nucleotides, so that you get two identical copies from one original

enzyme

Fig 30
The polymerase chain reaction

One cycle of the chain reaction takes a couple of minutes so, when repeated continuously, you can easily get over a million copies within an hour. There are sophisticated PCR machines that automate the whole process.

The polymerase chain reaction has many uses. For example, DNA samples taken at a crime scene can be copied within an hour to give enough DNA for forensic analysis. It can also be used in **genetic screening** as it allows us to take a tiny sample of tissue and multiply the DNA until there is enough to test for the presence of a particular disease. For example, in this way we can match a criminal to a crime scene from just one hair.

Enzymes and diagnosis

Phenylketonuria (PKU)

PKU is an inherited enzyme deficiency which affects about 1 in 10 000 people. In a healthy baby, the amino acid phenylalanine is converted into tyrosine by the enzyme *phenylalanine hydroxylase.* In cases of PKU this enzyme is missing, leading to a build-up of phenylalanine and other metabolic by-products. These substances interfere with brain development, so the condition must be diagnosed as soon after birth as possible. This allows treatment with a low phenylalanine diet before any permanent damage is done.

In the UK, all babies are tested for PKU soon after birth. In the **Guthrie test**, a sample of blood is taken – usually from the heel – and tested for phenylalanine.

Pancreatitis

The pancreas is an organ situated just under the stomach. It has two functions: to make the hormones (insulin and glucagon) and to make digestive juice. Pancreatitis is a severe inflammation of the pancreas, caused when the digestive enzymes begin to break down the tissues of the pancreas itself.

Pancreatitis is difficult to diagnose, because pain in the upper abdominal area could be the result of several different problems, e.g. gallstones, an ulcer, liver disease. Diagnosis involves testing blood samples for the presence of the digestive enzymes amylase and lipase. If they are present, it is a good indication that the patient has acute pancreatitis. This is because damaged pancreatic tissue is releasing digestive enzymes into the bloodstream instead of into the intestine.

Enzymes as analytical reagents

Enzymes are very sensitive and specific, so they are ideal for use in analytical tests. Enzymes can detect minute quantities of a particular substance, bringing about visible results in a short time. One example of this is the test for blood glucose levels. Diabetics need to assess the levels of glucose in their blood, so that they can match it to their insulin dosage and diet.

Glucose testing strips contain two enzymes (glucose oxidase and peroxidase) and a dye (DH_2), which changes colour when oxidised. The glucose oxidase catalyses the following reaction:

glucose + oxygen ⟶ gluconic acid and hydrogen peroxide.

The second enzyme, peroxidase, catalyses a second reaction:

DH_2 + hydrogen peroxide ⟶ water + D (coloured dye),

so the more glucose there is, the greater the colour change. The amount of glucose can be estimated by matching the colour of the test strip to a chart that indicates the glucose level for each possible colour.

12.9 Drugs are used in the control and treatment of disease

Beta blockers

Beta blockers are drugs used to treat people with heart disease and/or high blood pressure. They work by preventing the body from responding to adrenaline, the hormone that prepares the body for action by increasing heart rate and blood pressure and by initiating various other effects.

You may recall from Unit 1 that hormones work on certain **target cells**. The plasma membranes of these cells contain specific proteins into which the hormones fit. In this case, adrenaline fits into certain surface receptors of the heart cells, causing increased heart rate. Beta blockers fit into the adrenaline receptors, thus blocking adrenaline and preventing its effects.

Antibiotics

Antibiotics are a large group of synthetic and naturally occurring drugs that combat bacterial infection. Antibiotics work by interfering with prokaryote metabolism (Fig 31), leaving the eukaryotic cells of the host working normally. Basically, antibiotics work in one of three different ways:

- they prevent the formation of bacterial cell walls
- they interfere with DNA replication
- they interfere with protein synthesis.

D

***Bactericidal antibiotics** kill the targeted microorganisms.*

***Bacteriostatic antibiotics** prevent the targeted microorganisms from multiplying.*

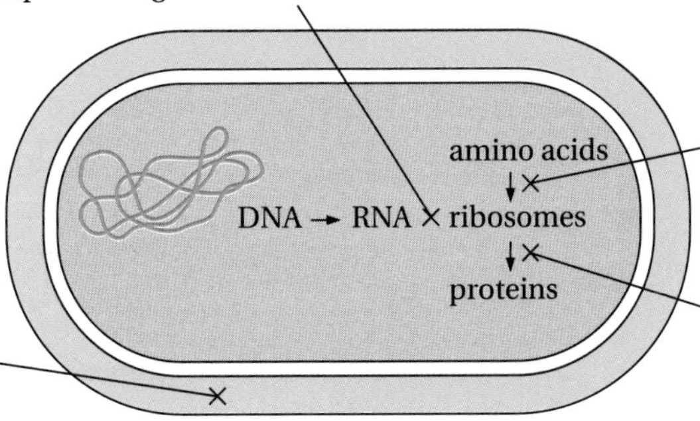

Fig 31
The mode of action of different antibiotics

Tetracycline is bacteriostatic. It binds to bacterial ribosomes, preventing RNA attachment

amino acids

DNA → RNA × ribosomes

proteins

Chloramphenicol is bacteriostatic. It prevents transfer of amino acids to ribosomes

Penicillin is bactericidal. It stops the formation of cross-bridges of peptidoglycan, an important molecule in bacterial cell walls. This weakens the cell walls. The bacteria absorb water and burst

Erythromycin is bacteriostatic. It binds to bacterial ribosomes and prevents translation

Monoclonal antibodies (MABs)

Antibodies have great potential in medicine because they are seen as 'magic bullets' – they are very specific and have the ability to target certain cells in the body, e.g. tumour cells. They could, in theory, carry toxic chemicals to exactly where they are needed, such as tumour cells, but this research is still in its early stages.

Monoclonal antibodies are so-called because they are made by a single clone of white cells. The cells that can make the required antibody are grown in culture, i.e. outside the body. They are stimulated to divide into a large clone that will then make useful amounts of the required antibody.

MABs have several uses:

- in medical diagnosis – MABs can be used to detect cancer cells, pathogens or even specific chemicals. MABs are used in pregnancy tests to detect the presence of the hormone **human chorionic gonadotrophin** (HCG)
- in cancer treatment – cancer cells have certain unique proteins on their cell surface. MABS can be attached to toxin molecules, which will then target these cancer proteins, delivering the toxin to its target with great precision
- in transplant surgery – MABs can be used to target and kill the cells responsible for the rejection of transplanted organs and tissues.

AS 3 Sample unit test

1 The diagram shows part of a molecule of DNA.

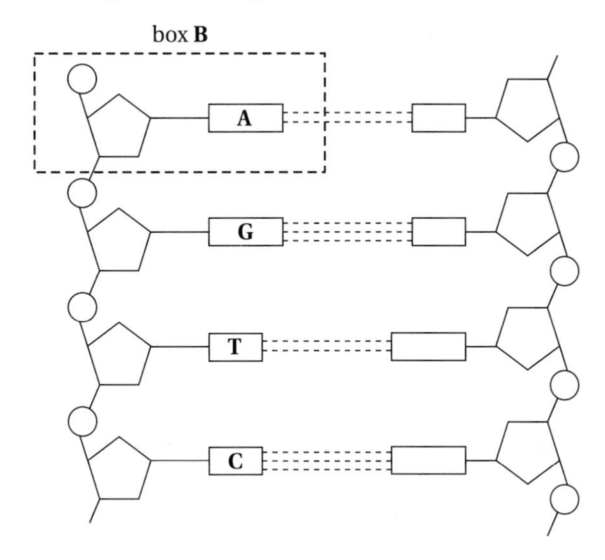

(a) Complete the diagram with the first letters of the appropriate complementary bases. *(1)*

(b) What is represented by:

(i) the part of the molecule in box **B**; *(1)*
(ii) the dotted lines between the complementary base pairs? *(1)*

(c) Describe how a molecule of DNA replicates to give two identical DNA molecules. *(3)*

(6)

2 The graph shows the growth of a population of bacteria over a five-day period.

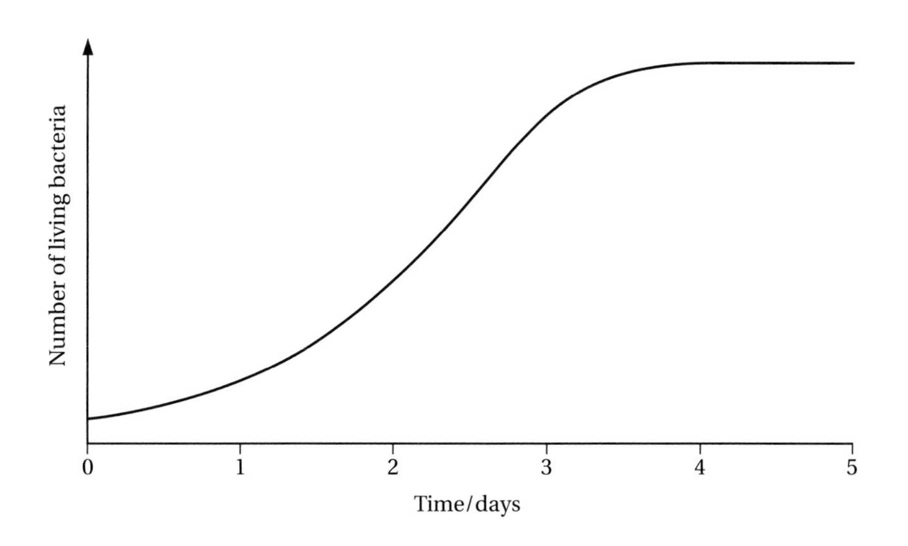

(a) Suggest an explanation for the shape of this curve after day 4. *(1)*

(b) This culture of bacteria was kept at a temperature of 20 °C. Draw a line on the graph to show the growth curve that would result from keeping the culture at a slightly higher temperature. *(2)*

(c) If an antibiotic had been added to the culture on day 2, how would you have known if it was:

(i) bacteriostatic; *(1)*
(ii) bactericidal? *(1)*

(5)

3 The drawing shows the structure of human immunodeficiency virus (HIV).

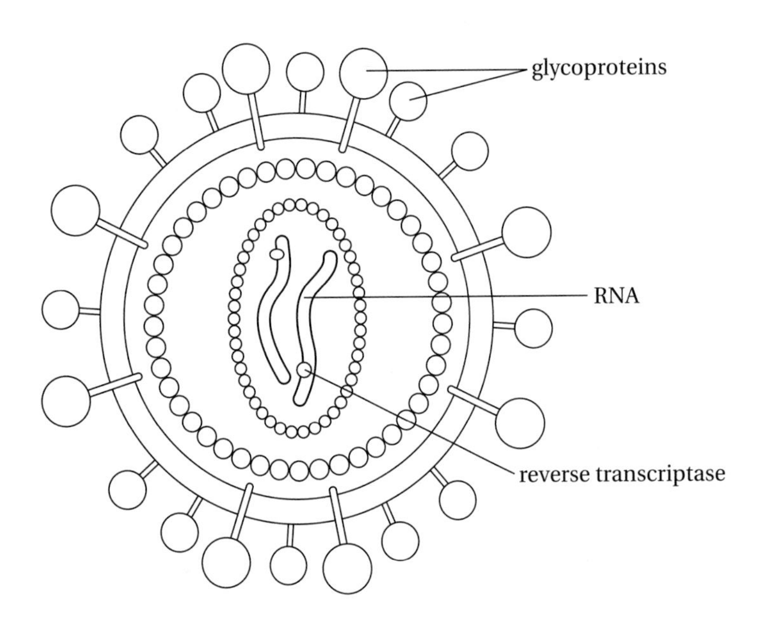

(a) The virus attaches to the plasma membrane of a helper T-lymphocyte. The virus RNA and the enzyme reverse transcriptase then enter the lymphocyte.

What is the function of each of the following:

(i) the RNA; *(1)*
(ii) reverse transcriptase? *(1)*

(b) One form of pneumonia and some types of cancer are normally very rare. They are much more common in people who are infected with HIV. Explain why. *(2)*

(c) Viruses such as the influenza virus and HIV have very high mutation rates. Suggest why this makes it difficult to produce a vaccine against them. *(2)*

(6)

4 The table shows some antibiotics and the way in which they work.

Antibiotic	Method of action
Penicillin	Prevents the formation of bacterial cell walls
Streptomycin	Distorts the shape of ribosomes in bacterial walls so that protein synthesis is stopped
Mitomycin C	Joins together the two polynucleotide chains which make a DNA molecule with strong chemical bonds so they cannot be separated

(a) Explain why:

(i) penicillin is effective against bacteria but not fungi; *(1)*
(ii) streptomycin cannot be used to treat diseases caused by viruses. *(1)*

(b) (i) Explain why cells treated with mitomycin C cannot synthesise proteins. *(2)*
(ii) Suggest why it is thought that mitomycin C might be effective in treating cancer. *(2)*

(6)

5 Plastic strips impregnated with certain chemicals may be dipped in urine. A change of colour indicates the presence of glucose. The test relies on the two chemical reactions shown by the equations.

$$\text{glucose} + \text{oxygen} \xrightarrow{\text{Enzyme } \mathbf{A}} \text{gluconic acid} + \text{hydrogen peroxide}$$

$$\text{blue dye} + \text{hydrogen peroxide} \xrightarrow{\text{Enzyme } \mathbf{B}} \text{green-brown dye} + \text{water}$$

(a) Name enzyme **A**. *(1)*

(b) With which of the substances shown in the equations are the plastic strips impregnated? *(1)*

(c) Explain why:

(i) the manufacturers recommend storing these strips in a cool place; *(1)*
(ii) the strips will only give a positive response to glucose. They will not give a reaction with other sugars. *(2)*

(5)

6 (a) The flow chart summarises one way by which tumours may arise.

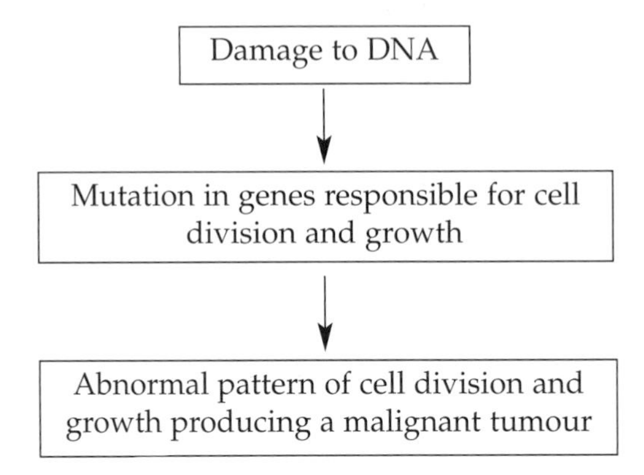

(i) Use the information in this flow chart to help explain why there has been a recent increase in skin cancer in the UK. *(2)*

(ii) Describe how a malignant tumour differs from a benign tumour. *(1)*

(b) Chemotherapy involves using drugs to kill tumour cells. The graph shows the effect of different doses of a drug used in chemotherapy on tumour cells and on healthy body cells.

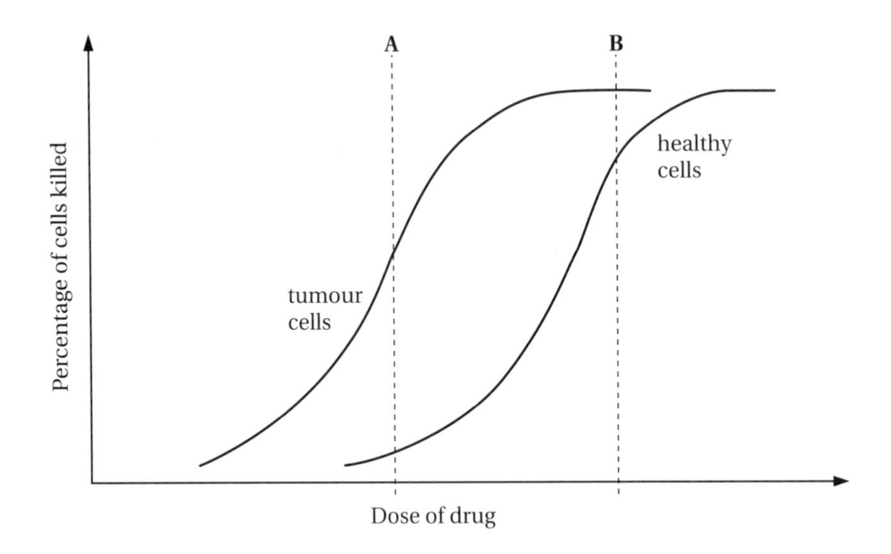

(i) Which dose of the drug, A or B, would you use treat a patient with a tumour? Give the reason for your answer. *(2)*

(ii) Explain why it might be necessary to give the patient several sessions of treatment with the drug. *(1)*

(6)

7 The table shows the results of some blood tests carried out on a patient admitted to hospital suffering from a suspected myocardial infarction (heart attack).

Substance	**Concentration in patient's blood/ arbitrary units**	**Range of concentration in blood of healthy individuals/ arbitrary units**
Urea	5.7	2.5–6.7
Cholesterol	8.2	3.6–6.7
Lactate dehydrogenase enzyme	2263	300–600
Potassium	4.3	3.4–5.2

(a) A myocardial infarction results in damage to the muscle of the heart.

(i) Explain how a blood clot may cause damage to the muscle of the heart. *(2)*

(ii) Lactate dehydrogenase is an enzyme found inside healthy heart muscle cells. Suggest why the concentration of this enzyme in the blood can be used to confirm that this patient had suffered a myocardial infarction. *(2)*

(b) Use the table to explain what is meant by a *risk factor*. *(2)*

(6)

8 (a) A microscope slide was prepared showing mitosis in a root tip from an onion bulb. Give a reason for carrying out each of the following steps:

(i) staining the tissue; *(1)*

(ii) pulling the stained material apart with needles before putting on a cover slip during mounting. *(1)*

(b) The graph shows the movements of chromosomes during mitosis. The curve shows the mean distance between the centromeres of the chromosomes and the corresponding pole of the spindle.

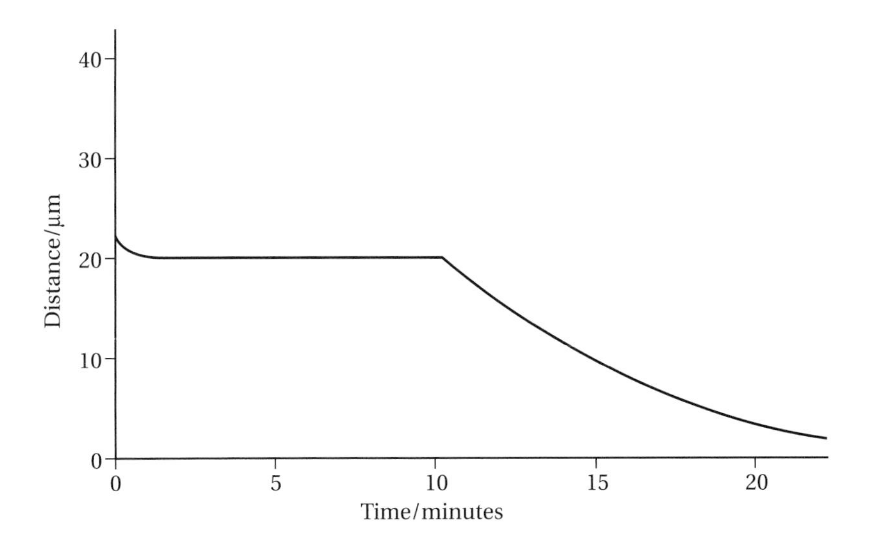

(i) At what time did anaphase begin? *(1)*
(ii) Explain how the graph supports your answer. *(2)*
(5)

9 A vaccine has recently been developed against malaria. A trial of this vaccine was carried out in South America. The graph shows some of the results of this trial.

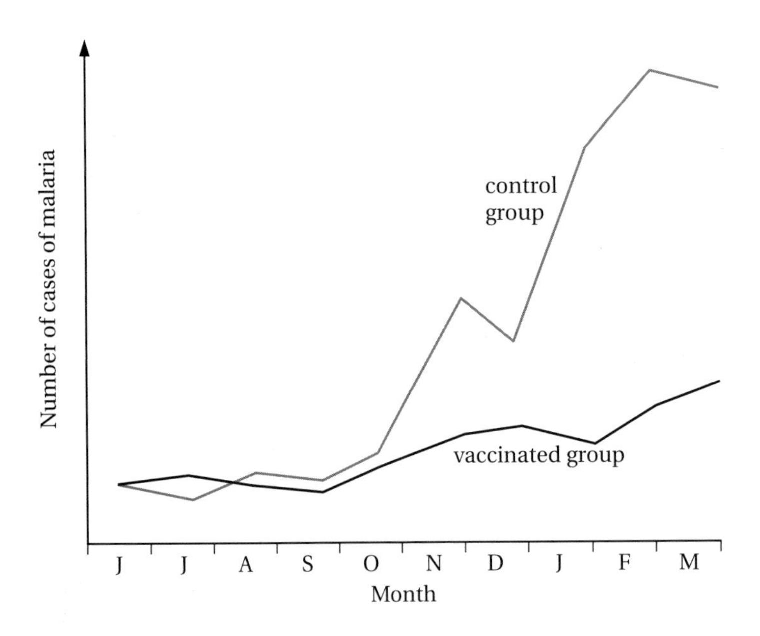

(a) (i) Suggest how the control group should have been treated. *(1)*
(ii) Explain why it was necessary to have a control group. *(1)*

(b) Describe and explain the evidence from the graph which suggests that there was a period of heavy rain from October to March. *(3)*

The table shows some more data collected during this trial. It shows the total number and percentage of people in different age groups who caught malaria during the first year of the trial.

Age group/ years	Vaccinated group		Control group	
	Total number	Percentage	Total number	Percentage
1–4	3	0.07	13	0.32
5–9	32	0.44	43	0.58
10–14	36	0.57	58	0.75
15–44	68	0.62	83	0.57

(c) Explain the advantage of giving the percentage of people who caught malaria as well as the total number. *(2)*

(d) From the data concerning the percentage of people catching malaria, the researchers concluded that the vaccine was most effective with people 1–4 years old.

(i) Explain the evidence from the table that supports this conclusion. (1)
(ii) Suggest why the vaccine was most effective with people in this age group. (2)

(e) Explain how B-lymphocytes, plasma cells and memory cells help to protect the body from disease. (5)

(15)

10 Read the following passage.

One aim of cancer therapy is to find a magic bullet that seeks out and kills tumour cells but leaves normal cells unharmed. For this to work, the bullet needs to be able to recognise a difference between the two types of cell.

Some tumours grow so fast that they outgrow their blood supply and the oxygen
5 concentration in their cells falls. Drugs are being developed that are only effective once they reach the low oxygen conditions inside a tumour cell. Here enzymes called reductase enzymes activate the drug which then kills the cell.

Professor Stratford and his colleagues at Manchester are taking advantage of the fact that the P450 reductase gene is only switched on in an environment which is low in oxygen. His
10 team have constructed the piece of DNA which is shown in the diagram.

Region of DNA which switches gene on in low oxygen concentrations	P450 reductase gene	Gene coding for protein which acts as a marker on plasma membrane

This piece of DNA was injected into breast cancer cells and the cells were grown in the laboratory. The marker protein was used to identify cells with the injected gene. When the oxygen concentration was reduced, the concentration of P450 reductase increased.

Use information from the passage and your knowledge to answer the following questions.

(a) Apart from the rates at which they grow, give one way in which tumour cells differ from normal cells. (1)

(b) Explain why the oxygen concentration in tumour cells may fall (lines 4–5). (1)

(c) Explain why the drugs mentioned in this passage do not kill normal cells. (2)

(d) Name the types of enzyme used to:

(i) remove the P450 reductase gene from a length of DNA; (1)
(ii) join the three pieces of DNA together. (1)

(e) (i) The investigators added the gene coding for the protein which acts as a marker on the plasma membrane to their specially constructed piece of DNA. Explain why. *(2)*

(ii) Some antibodies fluoresce when illuminated with ultraviolet light. Suggest why these antibodies could be used to identify the cells which had the marker protein on their plasma membranes. *(2)*

(f) Describe the parts played by mRNA and tRNA in producing a molecule of a protein such as P450 reductase. *(5)*

(15)

Unit test answers

1 (a) TCAG 1
(b) (i) nucleotide 1
(ii) hydrogen bonds 1
(c) chains part;
breaking of hydrogen bonds; semi-conservative replication/each chain acts as a template;
nucleotides/bases line up with complementary bases; join/polymerise 3 max

2 (a) accumulation of waste products/toxins/ depletion of nutrient supply 1
(b) curve rises more steeply;
flattens out to same final population 2
(c) (i) curve would have levelled out 1
(ii) curve would fall/number of live bacteria would fall 1

3 (a) (i) forms a genetic material that codes for virus proteins 1
(ii) makes a DNA copy of virus RNA 1
(b) normally controlled by immune system/T-cells;
people with HIV have damage to immune system/ T-cells 2
(c) always producing different proteins;
act as antigens;
antibodies only work against specific antigens 2 max

4 (a) (i) bacterial cell walls made of different substance to fungal cell walls 1
(ii) viruses do not have ribosomes/cannot synthesise proteins 1
(b) (i) two chains of DNA cannot be split apart;
in translation/to form mRNA molecule 2
(ii) cancer cells divide rapidly;
if more DNA is not formed will not divide/ cannot undergo mitosis 2

5 (a) glucose oxidase 1
(b) enzyme A, enzyme B and blue dye 1
(c) (i) enzymes not stable/denature at high temperatures 1
(ii) enzymes are very specific;
other sugars have molecules of a different shape;
will not fit active site of enzyme 2 max

6 (a) (i) increased exposure to ultraviolet light;
caused by decline in ozone/more foreign holidays/sunbathing;
damage DNA/increases rate of mutation 2 max
(ii) malignant tumour spreads and affects other organs 1
(b) (i) A as it kills a high proportion of cancer cells;
but has less effect on healthy cells;
so would cause fewer side effects 2 max
(ii) a single treatment would not kill all cancer cells 1

7 (a) (i) prevents blood flow to part of heart muscle;
insufficient oxygen 2
(ii) myocardial infarction results in damage to muscle cells;
enzyme released into blood when cells are damaged 2
(b) risk factor is a factor correlated with/thought to cause a particular condition;
cholesterol;
high cholesterol is correlated with/thought to cause heart disease 2 max

8 (a) (i) would show up chromosomes/nuclear content/genetic material 1
(ii) obtain a thin layer of cells 1
(b) (i) 10 minutes 1
(ii) distance between centromere and pole gets less/curve falls;
as chromosomes are pulled apart/spindle fibres shorten 2

9 Quality of Written Communication
Answers to part (e) of this question require continuous prose. Quality of written communication should be considered in crediting points in the marking scheme. In order to gain credit, answers must be expressed logically in clear, scientific terms.
(a) (i) injected with water/saline/something that did not include antigen 1
(ii) results could be compared with control group/ to make sure that nothing else in the treatment produced the results 1
(b) increase in malaria;
associated with increase in number of mosquitoes;
mosquitoes breed in wet conditions 3
(c) allows a comparison to be made;
as different numbers of people might have been treated 2
(d) (i) largest difference between the vaccinated and control group 1
(ii) they have not been exposed as much to malaria/fewer have had malaria;
no natural immunity 2
(e) B-lymphocytes respond to specific antigen;
divide rapidly/clone produced;
form plasma cells;
plasma cells secrete antibodies;
some form memory cells which become active on second exposure to antigen;
produce antibodies faster 5 max

10 (a) lower oxygen concentration/more reductase/ relatively large nucleus/difference in metabolism/ continue to divide when not in contact with other surface 1
(b) high respiratory rate associated with rapid growth and insufficient blood supply to supply enough oxygen 1
(c) they have enough oxygen present;
therefore reductase enzymes not produced;
inactive drug is harmless 2 max
(d) (i) restriction enzyme/endonuclease 1
(ii) ligase 1
(e) (i) the marker could be identified easily;
so it would be known which cells contain the piece of DNA 2

(ii) antibodies are specific;
proteins;
will only attach to marker protein;
correspond in shape 2 max

(f) mRNA makes a copy of the DNA code;
during transcription;
chains of DNA separate and mRNA nucleotides;
line up with complementary bases;
tRNA carries specific amino acid;
to correct position on mRNA;
anticodon and codon match 5 max

NOTES

NOTES

NOTES

Collins Support Materials for AQA – ORDER FORM

This booklet covers one module from the AQA Biology (A) course at AS-level.
If you would like to order further copies from the series, please send a completed copy of this page to Collins by fax or post.

Title	ISBN	Price	Evaluation copy	Order quantity
1 Molecules, Cells and Systems	000327706 2	£4.25		
2 Making Use of Biology	000327707 0	£4.25		
3 Pathogens and Disease	000327708 9	£4.25		
			TOTAL ORDER VALUE	

Also available:

Collins Advanced Science Series

Comprehensive textbooks for both years of A-level courses

Title	ISBN	Price	Evaluation copy	Order quantity
Biology	000322327 2	£24.99		
Human Biology	000329095 6	£24.99		
			TOTAL ORDER VALUE	

To order, please fill in your details and post to:
HarperCollins Publishers, Bishopbriggs, FREEPOST, GW2446, GLASGOW G64 1BR.

Name: ______________________

Address: ______________________

____

____

or

telephone on 0870 0100 442
e-mail us on Education@harpercollins.co.uk
fax your order on 0141 306 3750

For details of other A-level titles:
visit our website at www.**Collins**Education.com